Happy 78th.
With love
And

CHERRY'S
MODEL ENGINES

CHERRY'S MODEL ENGINES

The Story of the Remarkable Cherry Hill

David Carpenter

ROBERT HALE • LONDON

© David Carpenter 2014
First published in Great Britain 2014

ISBN 978-0-7198-1421-1

Robert Hale Limited
Clerkenwell House
Clerkenwell Green
London EC1R 0HT

www.halebooks.com

The right of David Carpenter to be identified as
author of this work has been asserted by him
in accordance with the Copyright, Designs and
Patents Act 1988

A catalogue record for this book is available
from the British Library

10 9 8 7 6 5 4 3

Design by Eurodesign
Printed by Craft Print International Ltd, Singapore

CONTENTS

	INTRODUCTION	7
CHAPTER 1	**EARLY DAYS** *A talent emerges*	11
CHAPTER 2	**HOW IT'S DONE** *'Secrets' of success*	19
CHAPTER 3	**EARLY SUCCESSES** *The start of something remarkable*	36
CHAPTER 4	**ON TO A NEW LEVEL** *Aveling & Porter road roller 1931*	45
CHAPTER 5	**AN INEVITABILITY** *Burrell showman's engine 1922*	50
CHAPTER 6	**FROM MAKER'S WORKS DRAWINGS** *Wallis & Steevens' Simplicity roller 1930*	58
CHAPTER 7	**FAITHFUL REPRODUCTION** *Wallis & Steevens' Advance roller 1936*	61
CHAPTER 8	**A SEMINAL MOMENT** *Taylor's Steam Elephant 1862*	66
CHAPTER 9	**THE JOYRIDE** *Savage fairground engines 1934*	73
CHAPTER 10	**RETURN TO VICTORIANA** *Batho 25-ton road roller 1870*	78
CHAPTER 11	**A MODEL MADE TO WORK** *Law & Downie road locomotive 1863*	87
CHAPTER 12	**PARISIAN INCLINATIONS** *Gellerat steamroller 1881*	94
CHAPTER 13	**THE GREAT PROJECT** *Andrew Barclay's traction engine and boring machine 1862–3*	99
CHAPTER 14	**A FAMILY AFFAIR** *Gilletts and Allatt traction engine 1862*	109
CHAPTER 15	**THE ENGINE IN A WHEEL** *Blackburn agricultural engine 1857*	117
CHAPTER 16	**FILLING IN THE DETAIL** *Blackburn agricultural engine 1863*	134
CHAPTER 17	**HOT COALS AND ICE** *Nathaniel Grew ice locomotive*	142
	MODELS OVERVIEW	147
	INDEX	148

INTRODUCTION

'It's not fair!' The comment came from a well-known model engineer talking to a friend at a Model Engineer Exhibition at Wembley in the late 1970s. He was standing in front of Miss Cherry Hinds' latest model, carefully displayed at eye level.

His friend remained quiet for a while then said, 'You know, there are several thousand blokes here, most of them experienced engineers, and not one of them can produce work like that.'

That, in a nutshell, sums up the quality of the work of Cherry Hill, as she is now. But there is more to it than gifted craftswomanship. All of her models have that something special that makes each one stand out, that draw exhibition visitors to them like a magnet even in a room filled with excellent models.

In no small part that is due to the selection of what to model. Most model engineers will choose things that are familiar, from publications, websites, and so on, or even from fondly remembered encounters. Cherry also started in that way making models of steam engines, traction engines and fire engines that were familiar, at least as preserved items. But her model-making career took a different turn. It moved into the Victorian era, with its engineering flights of fancy. The creativity of those old engineers was matched by that of the daughter of an agricultural machinery engineer from Worcestershire who had the ability to turn drawings, often incomplete, into working models that sometimes could not have worked if made to their original designs.

So, rather than the familiar traction engine models, Cherry's are unusual. One, for example, has the boiler fixed inside the vehicle's single driving wheel.

Overleaf Cherry Hill receives the Duke of Edinburgh Award from Chief Judge, Ivan Law, at the Centenary Model Engineer Exhibition in 2007.

A Hill model combines extraordinary craft skills with a mind that will delve for months into research and then produce a working design. This book sets out to record these models and to explain the processes used to produce them. Also, to show the models in a way which is revealing of the work that goes into them, particularly through the photographs of Cherry and Norman Mays (who photographed all of the early models).

There is no temptation here to make the descriptions anything more than additions to the images, so the superlatives will be few and far between because they are unnecessary. It is not sensible to add hyperbole to what is genuinely perfection.

Thus it is the images of the models and how they are made which dominate, together with the remarkable picture by Michael Jones, former deputy editor of *Model Engineer* magazine, of the hands that do the work.

Once completed and shown, all the models have been given away. Cherry is reticent about her success. However, we have to record the recognition she and her models have received.

In 1989 and 1995 Cherry was awarded the Sir Henry Royce Trophy for the Pursuit of Excellence. In 2000 she was made an MBE for services to model engineering. Then, in 2004, she was elected a Companion of the Institution of Mechanical Engineers. In the same year Cherry was made an Honorary Member of the Society of Model and Experimental Engineers.

At the annual Model Engineer Exhibition in London, Cherry has won nine Gold Medals and is nine times winner of the Duke of Edinburgh Award, the highest accolade for engineered models.

She has also won the Bradbury Winter Memorial Trophy eight times. Other awards include the Aveling Barford Trophy, the Crebbin Memorial Trophy, and three Championship Cups (the Championship Cup was the forerunner of the Gold Medal and only one was awarded each year). The major awards were all for the later models, but the early models all received awards, too.

The hands that do the work holding a Blackburn engine cylinder that was scrapped because one stud boss was fractionally too long. *(Michael Jones)*

CHERRY'S MODEL ENGINES

Gold Medals, Royce Awards and the Duke of Edinburgh Award, the highest accolade in model engineering now won nine times by Cherry.

CHAPTER ONE

EARLY DAYS

A talent emerges

Cherry at the workbench in the Malvern workshop. The bench had been her father's.

Overleaf Cherry Hinds with her first traction engine, the Allchin Royal Chester in 1:16 scale.

Cherry Hinds was born in Malvern in 1931, the second of three daughters. Her parents were originally from Kent, her father, George, having been born in 1900 at Gilletts, the family farm in Smarden, Kent. The name Gilletts will return in the description of one of Cherry's engines, as will her mother's maiden name, Allatt. The farm was sold in 1926, and George moved to Malvern the following year, where he sold fertilizers. He also started to build up a workshop, including a 1914 Pittler lathe that had been used to turn cartridge cases for World War I. In 1936 he was able to start an engineering business in Worcester, making hop-picking machines.

At the outbreak of World War II the government closed the business. As machinery was in short supply and needed by the War Office, George Hinds offered his

CHERRY'S MODEL ENGINES

workshop to them, and he went along with it to the Royal Aircraft Establishment at Farnborough. It returned to Malvern in 1946.

Cherry was seven years old when the war broke out. She clearly remembers the workshop as it was, with its framework for overhead shafts, fast and loose pulleys, the workbench (which is still her main bench), a tall chest of drawers acting as a drill stand (as it still does) and the Pittler lathe (which she still has). 'My earliest memory is of a small drill left in the drill chuck. I could reach the lever by standing on the waste box. Not realizing why operating the lever would not drill a hole in a piece of wood, I was not bright enough to think the drill should be turning!'

The first thing Cherry made, aged about nine or ten, was a small chest of drawers about 8 x 6 x 8in. Eventually everything fitted satisfactorily until after painting. The drawers then went in, with difficulty, but would not come out again. Next, aged eleven, was a scooter made for her younger sister. While it looked sturdy and good, someone suggested it would be better if it was jointed and made steerable.

'So that was attended to. Nuts, bolts and coach bolts from a scrap box left behind when the workshop went. I learned about Whitworth and BSF threads, and that sometimes it took a very long spanner and generated a lot of heat to make some of them fit together.'

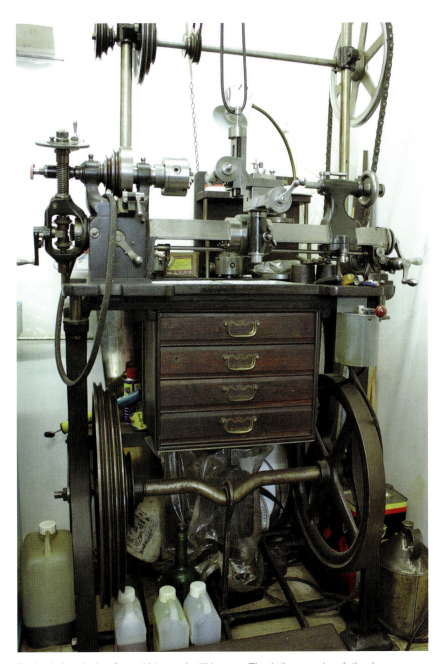

Pittler lathe dating from 1914 and still in use. The lathe was her father's.

Husband Ivor at home in Florida working at his Clausing lathe.

During the war Cherry won second prize in a scratch-built model aircraft competition with a Sunderland flying boat, and received a special mention for fretted-out propeller blades. As with most children, her model-making during the war involved all types of aircraft and a fleet of warships.

Cherry was sent away to boarding school during the war, initially in Kent, then the school was evacuated to the north of England, returning to Kent in 1947. 'I saved to buy myself a second-hand two-stroke Excelsior 125cc motor bike at the age of sixteen. This was replaced later by a four-stroke Matchless 250cc. Parental permission was difficult to obtain!'

After school, Cherry went to St Andrews University before joining the family firm. 'I joined at a lowly position and was sent out for the ice creams, etc. I worked in the works and, as the firm grew, a drawing office was created to design the hop-picking machines.'

George Hinds died in 1961 and the company was bought by the Wolseley Hughes Group in 1968. Cherry went to work for Bruff Engineering in the same group – the only other company making hop-picking machinery in the UK at the time. She was made redundant in 1975 and went to work at Autopack, a weighing and packaging business. She remained there working in research and development until she left in 1984 to get married to Ivor Hill.

Although born in England, Ivor has lived in the United States since he was a boy. He is a professional toolmaker and his work is also his hobby. By the 1960s Cherry was becoming noticed as a very good model engineer, and her work featured in magazine articles. After one exhibition Cherry appeared on the front cover of *Model Engineer*. Ivor saw it and was captivated: 'I'm going to marry that girl', he vowed. Some years later, he did.

Ivor engineered a meeting shortly after seeing that photograph. From that early time he wanted to help Cherry's model engineering activities as well as persuading her to marry him. He realized she would not accept gifts, so he asked if he could 'park' a new Boxford lathe at her workshop. (That lathe eventually went to her brother-in-law and hot air engine builder, Norris Bomford, when a Myford Super Seven arrived in 1988. That was, in turn, replaced by a Myford Connoisseur in 2004.)

Until Ivor and Cherry married in the eighties, they remained separated by the Atlantic Ocean, meeting up when visiting model engineering shows in the UK. In recent years, Ivor has stopped travelling to the UK with Cherry to visit exhibitions in London and Harrogate but they spend half an hour every evening on the phone to each other. Ivor retains his workshop at their home in Florida, and Cherry hers at their home in Worcestershire.

Engineering had really taken a hold during the university vacations when Cherry set about building a Humber special car. The basis was a 1926 8 hp Humber chassis with engine axles and wheels, and steering box, column and wheel. The chassis was reduced in width by 8in and in length by 26in. Austin 7 axles and wheels were fitted. The engine was retained, along with the gearbox, steering (with column lowered) and suspension. Austin spider-type shock-absorbers were fitted. Cherry made her own hydraulic brakes. The body work was duralumin on a wooden frame, with a hinged

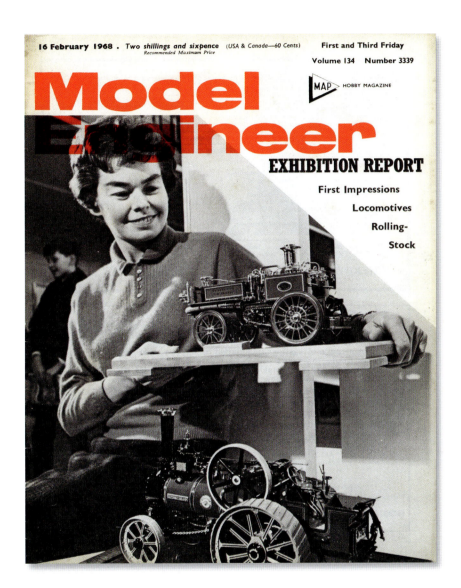

16 February 1968, cover of Model Engineer magazine. Ivor vowed to marry Cherry after seeing it.

CHERRY'S MODEL ENGINES

bonnet and cycle-type wings. The car was roadworthy and registered for the road in 1952. However, it was quickly sold and replaced with a 1936 MG TA.

A taste for sports cars continued, with a Triumph TR2 in 1955, a Jensen 541 in 1961 and a 541R in 1963. The Jensen 541R was the dream car of every young man of that time, rather like a Ferrari today. After that 'model engineering proved a more satisfying hobby than wax polishing a black motor car', and the 541R was replaced by an Austin A40 in the late 1960s.

The Jensen had suffered minor damage while Cherry was visiting a hop-growing cousin. She literally bumped into him in a country lane, slightly damaging the radiator. The local branch of Sercks, the radiator makers and repairers, only worked on detached radiators. They were politely amused when this young lady asked if she could park outside and remove it. They did carry it into the workshop and offered to refit it after repair. The offer was refused, as was the offer of the magic Swarfega. ('I always carried my own.')

Some years after selling the Humber, Cherry was surprised to hear from a local farmer who wanted to rebuild it to its original form, and asked whether she knew the whereabouts of the rear axle, wheel hubs and most connections to the back of the gearbox. He was delighted that she still had them and doubly so when they were offered in exchange for a large pumpkin. Unfortunately, he could not find a pumpkin but sealed the trade with a large mangle-wurzel and a very large box of York chocolates.

In the 1950s, Cherry designed a pair of electric scissors, with dressmakers in mind. Based on a motor similar to that of an electric shaver it was successful enough to be granted a patent. However, after three years she was told that a Frenchman had beaten her to it in 1918.

Opposite Cherry with the Allchin and Merryweather fire engine in July 1968.

Prototypes of the Crypton Synchro-check carburettor balancer designed by Cherry and taken up by manufacturers AC Delco.

Cherry setting up her small Cowells lathe in the workshop.

Another patented piece of equipment was a carburettor balancer, to equalize the intake of SU carburettors fitted in pairs, or more. Cherry designed and built the prototypes and a friend looked after the business side – patent, dealing with the makers (AC Delco) and collecting the money. It was called the Crypton Synchro-check and was popular among sporting car enthusiasts and manufactured for around eight years during the 1950s. Examples are still sought by classic car enthusiasts.

For all the success with full-size engineering, it was to be model engineering which took over Cherry's life, and that all began with a copy of the Bonds O' Euston Road catalogue which had always been in the Hinds' workshop. In 1953 Cherry had the chance to visit them for some small pieces of brass. She also left with a set of castings for the Stuart No. 9 steam engine, bought as a starter model, after the ME Traction Engine had been dismissed as 'too expensive and too difficult'. The rest, as they say, is history – and still counting, with some eighteen models now completed, all of them outstanding, representing thousands of hours of work each. And all given away once completed!

No doubt, there are more in the pipeline.

CHAPTER TWO

HOW IT'S DONE

'Secrets' of success

With the early engines, Cherry worked with designs that existed as models, and/or in full size. However, these exercises, which were mainly workshop based, gave way to completely original work which had never been modelled before and, in some cases, not built in full size, either.

With thousands of engine designs held in archives, rarely seen for more than a hundred years, which ones to choose? Likely candidates were found after extensive burrowing in old magazines, and a final choice made on the basis that it had to be an unusual design, a challenge to make, and capable of being made to work. It is this last criterion which separates Cherry's models from most top examples of models found in museums.

The next step in the process is to try to visualize how it will look when it is completed about seven or eight years later. That is how long it takes. Each new model goes through many phases. Research, preparing initial dimensions, sometimes making a mock-up model, preparing working drawings, construction, testing, photography, painting and, finally, more photography. Early photography was done professionally by Norman Mays, but Cherry has done her own photography since 1986, initially with film, but she has since 'gone digital'. Later engines have also been captured on camcorder when they were tested, running around the garage floor powered by compressed air. This method has been used to test all the engines apart from the Burrell and the Advance roller, which were tested on steam.

Most model engineers spend most of their time in the workshop, machining and assembling components. Cherry spends less than

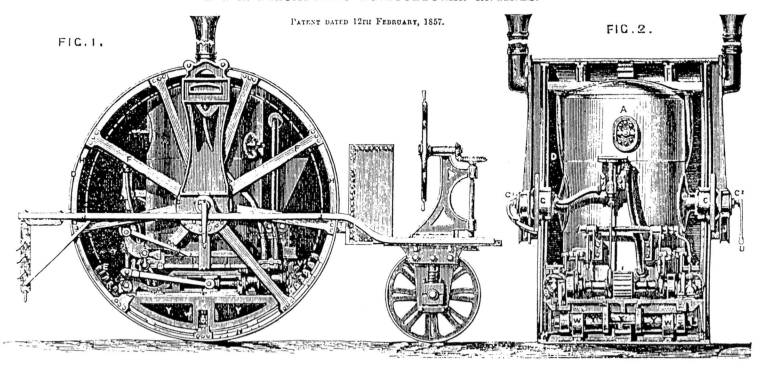

I. & R. BLACKBURN'S AGRICULTURAL ENGINES.
PATENT DATED 12TH FEBRUARY, 1857.

Models for the Blackburn Agricultural Engine of 1857 were carefully researched using patents and contemporaneous journals.

half her time in the workshop, with the majority being spent on the other aspects of a project.

Finding the right engine to model comes from combing through early volumes of *The Engineer*, *Engineering* and *Mechanics Magazine*. Here there are plenty of 'interesting' old steam rollers, road locomotives and 'traction engines'. From these, the ones that are also 'unusual' are selected for further investigation.

This (novel) invention consists in a method of constructing engines to be employed in agriculture in which the boiler and steam cylinders are suspended inside of a drum of considerable diameter, so that they shall always be in the same plane, while they cause the drum through toothed gear to rotate round them, and thus perform an onward motion. From the extended bearing of the engine, and the easy rolling motion imparted to it, it will advance steadily over rough and uneven ground, and may be employed to drag after it ploughs and other implements; or implements may be driven by it in advance. By itself it may be employed as a roller, or as a means of transporting matters over common roads. The invention further consists in mounting a cylinder or cylinders and boiler upon a frame supported by two or more drums, which are caused to rotate by gear communicating directly with the outside of the drums. The drum or cylinder which is employed is made of malleable or cast iron, combined or not with wood. On the two outside rims of the drum are fitted iron spoked wheels with hollow cast-iron centres or naves firmly bolted or rivetted to the drum. Within the drum are placed the boiler and engines, which are suspended by trunnions or flanged axles rivetted to the centre of the boiler, and which axles fit into the hollow centres or knaves of the outside wheels of the drum. The boiler and engines are thus evenly suspended within the drum. To the inside of the drum is bolted internal spur or other gearing, which is set in motion by a pinion or other suitable agent connected to the crank shaft of the engines. In order to prevent the boiler and engines from oscillating, or changing their relative and vertical positions, there is rigidly keyed on to the trunnions or axles on the outside of the drum, a frame of iron, or other suitable material, which is carried forward on both sides beyond the circumference of the drum, and to one or both ends of the frame are attached wheels or drums of small dimensions fixed on a shaft or shafts, which serve the double purpose of guiding the locomotive and also to receive the strain of the engines while at work, and thus become the medium of resistance between the engines and the earth. These wheels or drums, now alluded to, are fastened on to an axle or axles resting in journals on the outside frame above mentioned, and are, by bevel wheels, shafts, and hand wheel, so inclined by manual power to either side, that they cause the outside frame to act as a lever, and this frame being rigidly keyed to the axles of the boiler, guides and turns the locomotive at pleasure. Upon the frame, either before or behind the large drum, is placed a tank, holding water, which is supplied to the feed pump of the engines through one of the axles which suspend the boiler and engines, such axle being hollow through the greater portion of its length. The axle on the opposite side is also hollow through the whole of its length, and into it is placed a spindle, communicating with the steam regulator in the inside of the boiler, by which the engine driver is enabled to regulate the supply of steam, and have control over the engines while the machine is in motion. The boiler is furnished with a fire feeding apparatus, with a large hopper to hold fuel, which is set in motion by the gearing of the drum, thus enabling the engines to work for two or three consecutive hours without stopping to feed the furnace. Attached to the engine frame at the bottom of the boiler are friction wheels, which revolve on the rail iron forming part of the shell or ribs of the drum; these wheels take part of the weight of the boiler and engines, thus assisting the two outside bearing wheels of the drum to carry the boiler and engines and give steadiness to the engines while at work.

Fig. 1 represents a side elevation, partly in section, of an apparatus constructed according to this invention; and Fig. 2, a front view of the same. A is a multitubular steam boiler; B, B, are a pair of horizontal steam engines, coupled; C^1, C^2, are the axles or trunnions, which are supported in the ends of the drum D; E is an internal spur wheel, into the teeth of which the spur pinion I works; F, F, are bearing wheels, formed with iron spokes, and bolted to the rims of the drum; G, G, are cast-iron centres of the bearing wheels; H, H, are the ribs of the drum formed of T iron; J, J, are outside frames attached to the guiding apparatus and to the drag bars; K is one of four wheels of cast-iron, mounted on a shaft, and carrying plummer blocks at each end, attached by brackets to a circular plate L, which has teeth cast round one-third of its circumference, and a centre pin hole with a slot on each side, into which guide pins traverse, which pins are firmly bolted into the cross plate; M, M, is a cast-iron plate, bolted to the outside frames, containing three pins for the circular toothed plate L, also forming the foundation plate for the guiding bracket carrying the shafts and wheels P, Q, R; N is a pinion, on a vertical shaft or spindle, having at the upper end a bevel wheel O keyed on to it, which bevel wheel is driven by a smaller bevel pinion P, keyed on a short shaft, turned by the hand wheel R on the other end of it; S, the water tank and tool box; T, the water pipe, connected with the pump through the trunnion C^1; U is a handle at the end of a spindle, bearing in the trunnion C^2, connected with a steam regulator within the boiler for supplying the engines with steam; V is a fire-feeding apparatus, with coal hopper, driven by pinion V^1; W, W, are friction wheels, bearing on the rail irons X, X; Y, Y, is an ash pan; Z is a draw bar, furnished with links and a rack. It is proposed also to construct an engine similar to the above, but fitted with tyres, so as to adapt it to railways.

In the early 1990s just eight were selected as potential model projects. For those selected, a search was made of patentees in the relevant years at the London Science Library at Southampton Buildings in London, now sadly demolished. Some of them were not listed. That was followed by a search through the 'patent abridgments' under several sections,

HOW IT'S DONE

The finished model Blackburn, the result of six and a half years work involving 5,200 individual parts.

23

Right Elephant engine mock-up. Many parts are built to final model standards, but not used in the finished model.

Far right Law & Downie engine mock-up made from brass, wood and dural.

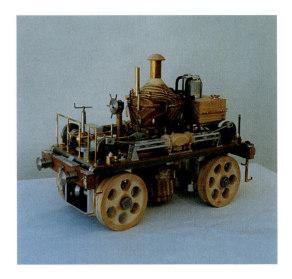

throwing up more names. That led to a search through their old patents. In those days it was possible to search through the old leather-bound volumes in the library vaults. Many visits were made over the years for lengthy searches, which often revealed nothing more than outline details. Importantly, however, it was possible to obtain prints of original outline material.

All that information was stored and provided the source for selection of the next project. Decisions were not hurried. They laid down the work programme for the next seven or eight years.

Once the decision was made it was back to the Science Library to search for anything that might have been missed earlier and to extend the research to include correspondence and magazine articles, plus particulars and history of the builder. At this stage, depending on the information available from the patent, some initial drawings were prepared and, if necessary, a mock-up made from wood, brass and dural to check the initial sketches. Mock-ups of the Taylor Elephant and Law & Downie engine are shown in the adjacent photographs.

Detailed design and full drawings followed, with workshop construction of the model begun as the drawings progressed.

What are the secrets of actually making these wonderful creations? There aren't any, really. Everything, and that really is everything, is made by conventional machining and hand work in a well-equipped workshop. There is no CNC. At the time of writing, Cherry has yet to get around to owning a computer.

The 'secret' of success is that everything

Research and design are the most challenging elements of a Cherry Hill model.

has to be spot on. For example, a cylinder made for the Blackburn engine had a stud boss 1mm too long, so it was scrapped.

'But, Cherry, no one could possibly have known.'

'I would!'

The house in Worcestershire was built with a workshop very much in mind. There are two semi-basement rooms for the machinery and workbenches, a drawing office containing a large draughtsman's parallel motion drawing board. A separate room at the end of the garage houses a surface grinder and the paint booth.

The main equipment in the workshop includes several milling machines – Myford, Emco and Centec – and a number of lathes. The main lathe is a Myford Connoisseur. There is also an IME watchmaker's lathe, and a very accurate Cowells lathe for small work. There is also the first lathe from her father, a treadle-operated pre-World War 1 Pittler lathe which is still in use, now permanently set up for dividing work.

The Myford milling machine is the largest machine in the workshop.

The Centec is a well-used milling machine.

HOW IT'S DONE

The Connoisseur lathe from the late, lamented Myford factory is the favourite.

There are several drilling machines including a Startrite, a pre-World War II Gamages sensitive drilling machine and a high-speed drilling machine. Other well-used machines include an off-hand grinder, polisher and bandsaw. At the other end of the spectrum, there is a row of tiny home-made BA spanners going down to 16BA increasing upwards in gaps of 0.002in to cover any eventuality. Other home-made accessories include the division plates drilled on their edges shown in the photograph overleaf.

Home-made small spanners go down in stages of 0.002in to 16BA.

Division plates drilled around the edges made in the workshop.

Cherry and Ivor have another well-equipped workshop at their home in Florida which, to Cherry's delight, includes a planing machine. This, at 120 years of age, is even older than Cherry's Pittler lathe. Other major items of equipment include a Clausing lathe and a milling machine from the same manufacturer.

The workshop in England is kept neat, with storage wonderfully organized. Even matchboxes are brought into use and fitted with dividers, as in the photograph overleaf, to hold tiny screws and other components. With thousands of minute components in these models, good organization is essential.

Ivor's Clausing lathe.

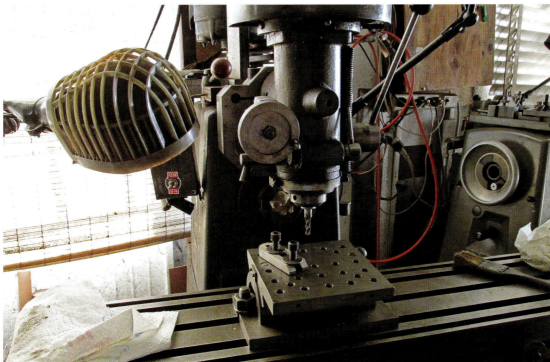

Ivor's Clausing milling machine.

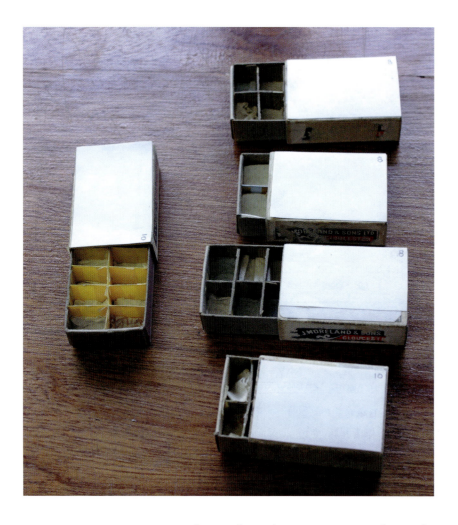

Hundreds of tiny parts need careful storage, so divided matchboxes are used.

Machining brain-teaser – set of symmetrical polyhedra made in brass without a computer.

It is making these tiny parts in the workshop rather than buying in ready-made versions that adds to the length of time a Cherry Hill model takes to complete. They also add to the search for perfection so characteristic of everything she does. Tiny rivets are turned from steel bar, chain links are individually made to create small chains, and all the screws (apart from a few used in construction and invisible on the final model) are made in the workshop. Important threads for glands, unions, studs and so on are screw-cut on a lathe with ME thread forms. Smaller threads are cut with taps and dies and these go down to 16BA. Even smaller threads are made to metric standards.

A whimsy was to make a set of symmetrical polyhedra out of brass. (Anyone looking for a machining brain-teaser might like to try to make the set comprising tetrahedron, hexahedron, octahedron, dodecahedron and icosahedron.) That set lives in a small cabinet with other items from times past, including a lighter made for Cherry's father, and the prototype carburettor tuning device taken up commercially by AC Delco.

Nothing would be possible without well-organized materials.

People often ask about the workshop techniques used. Generally speaking they are the same conventional techniques used in workshops around the world. However, they are chosen to produce a perfect result, not to save time. Cherry does make extensive use of specially made jigs and fixtures for holding and positioning parts while they are being machined (see Chapter 15 for examples). With so many small and irregularly shaped parts involved in these models the extra work involved does pay dividends. Indeed, often there is no alternative.

Some of the parts are so small that it is not possible to make them as working items. Pressure gauges and injectors are dummies where the pipework is below ¹⁄₁₆in. Similarly all the very small steam cocks are dummies, and lamps are non-working with the exception of the Burrell showman's engine which is built in a larger scale. All parts are made from metal bar. Components which were castings on the prototype are fabricated on the model. Castings were only used on the two early projects to Stuart Turner designs, and for the cylinders and wheel hubs for the Burrell showman's engine, for which Cherry supplied patterns and materials to the foundry.

Making virtually everything on a model is demanding in all sorts of ways, and new skills have to be learned. For example, as the models made became more demanding, producing nameplates could only be done well by photo-etching rather than cutting letters out of thin brass sheet. The easy answer is to contract such jobs out to specialists, but Cherry took six months to learn how to carry out the process and to make the kit needed.

Once all the parts are made and photographed, they are assembled and photographed again. At this unpainted stage a display case is made and the model is first shown at the annual Model Engineer Exhibition. This is an excellent opportunity to study the craftswomanship involved, and the quality of machining and hand finishing.

At around this time, work usually starts on the next engine. That will already have been 90 per cent researched, and the process starts again. Meanwhile, the recently completed model is stripped down and all the parts painted individually. Sub-assemblies are put back together, and the whole thing is finally assembled and fitted in its case to go back to the exhibition to be judged. There are times when three engines might overlap; one being worked on, one ready for dismantling and painting and another being researched.

All of the fine work on a model can be ruined if it is painted badly. Cherry's models have all been beautifully finished. It is interesting, and helpful to others, to see how her technique has developed to achieve those exceptional finishes.

All models are painted with cellulose paint. The first engine, the Stuart No. 9, was given three coats of undercoat and three top coats, applied with a brush and rubbed down between coats. The next three engines were sprayed using a Grafo air brush. Cherry first asked the manufacturer if it was all right to use cellulose in the air brush. The answer: 'Don't know! Try it and see!' Clearly it worked. Bronze and brass parts were acid-etched prior to painting. The paint used was thinned to a ratio of 1:3 of thinner. Coats were not rubbed down, to preserve the fine definition. Four or five very thin layers of undercoat were followed by a similar application of top coat. Top coats can be sprayed as soon as the previous coat is dry – in less than five minutes. However, it is best to leave a day or thirty-six hours between undercoat and the first top coat. The Grafo gave way to a Badger air brush and that make has remained in the paint shop. Cherry makes use of all three sizes of head assembly and needle, depending on the part being painted.

All the early models, up to the Wallis & Steevens steamrollers in 1979, were painted entirely in gloss paint. Lining was done using gouache artist's water-based paint applied with a fine sable brush, and fixed with three sprayed coats of clear cellulose blending. When two or more lines and colours were needed the three coats of blending were applied between each. Masking tape, of a type that does not allow the cellulose to 'creep' under its edges, was used to mask off the different colours. Other methods of lining have been used, with Humbrol enamel thinned and applied with the same brush as the gouache to rule the lines. Talcum powder is used as a matting agent when required. The same paint has also been used with masking each side for the fine line, enabling a larger brush to be used, with a hand that does not need to be as steady.

From 1982 and the Tayor's Steam Elephant model, matt paint has also been used. Different shades are obtained by adding in varying quantities of suitable matting agent to gloss paint. Taylor's Elephant is a black engine with no fewer than seven shades of black paint used to produce the desired effect.

Cherry's techniques are not unique. What makes her models different is the care and finely honed skills brought to bear in that search for perfection. For all the workshop skill, it is the research and design elements of a Cherry Hill model that separates it from the run of the mill.

Views of the Barclay engine (see Chapter 13) in bare metal and painted.

HOW IT'S DONE

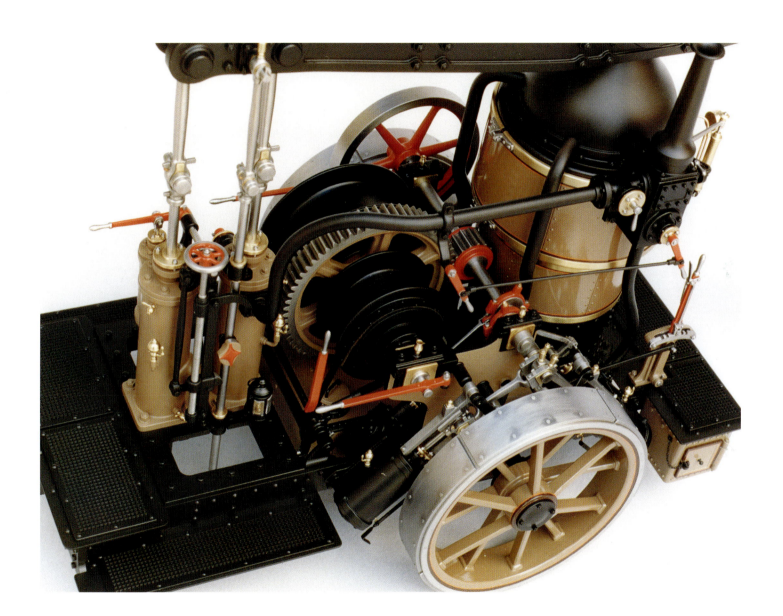

CHAPTER THREE

EARLY SUCCESSES

The start of something remarkable

Cherry Hinds' first model, like that of many another budding model engineer, was sourced from Stuart Turner (now Stuart Models) and bought as a set of basic castings from Bonds O'Euston Road, a favourite haunt of model engineers of the day. It is a small horizontal steam engine, the Stuart No. 9, widely regarded as one of the most attractive engines to grace the Stuart Turner range. It is not a model of a specific prototype, but a typical large, powerful single-cylinder engine of the type used to power machinery in mills and factories before electric power took over.

It took her eighteen months to complete between 1956 and 1957. When Cherry completed the basic engine she was 'thrilled to bits' when it burst into life after being plugged into a compressed air line: 'For the first time I experienced the amazement that I had managed to build an engine that ran!'

Encouraged by that, she ordered castings for the accompanying pump and governor.

The Stuart No. 9 measures 11 x 6¼ x 5¼in, and resides in the wooden-framed case she built for it.

It was a sign of things to come. For a beginner it was very well made, earning a Bronze Medal at the 1964 Model Engineer Exhibition at a London hotel. She also improved the Stuart design somewhat, to give it a more 'scaled' appearance by using smaller nuts, for example, than those specified by Stuart Turner. And so the die was cast, for the next increasingly fascinating and exquisite models that emerged from her workshop.

The 9H was given to the Society of Model and Experimental Engineers (SMEE) where it remains, occasionally given an outing to exhibitions around the country, where it is

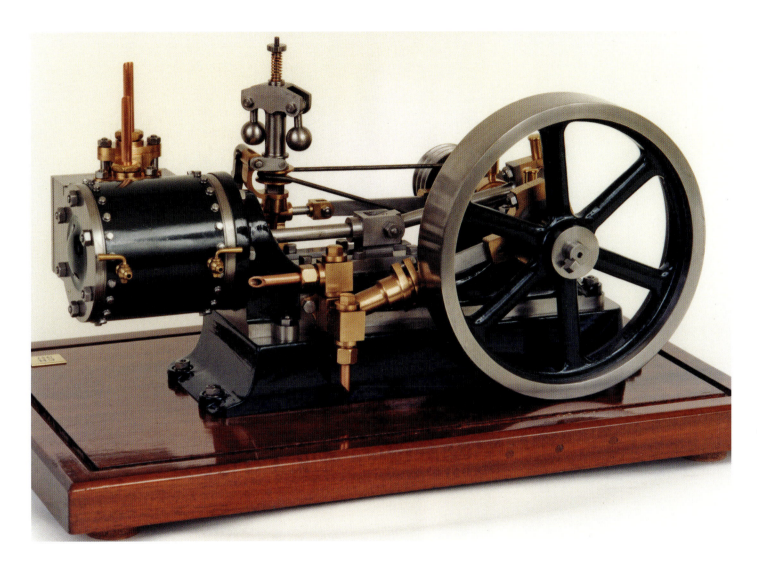

interesting to watch people who have no idea what it is, stop and realize it is something a bit special. The SMEE has also played an important part in Cherry's life. Both she and husband Ivor are both long-standing, active and enthusiastic members.

The 9H is still available from Stuart Models as a set of un-machined castings, materials, fixings and drawings. It's not a bad way to start in model engineering.

After completing the relatively simple steam engine, what would be the next model?

The very first model, the Stuart No. 9, incorporated improvements by the novice maker.

The Allchin Royal Chester traction engine – the first of many.

EARLY SUCCESSES

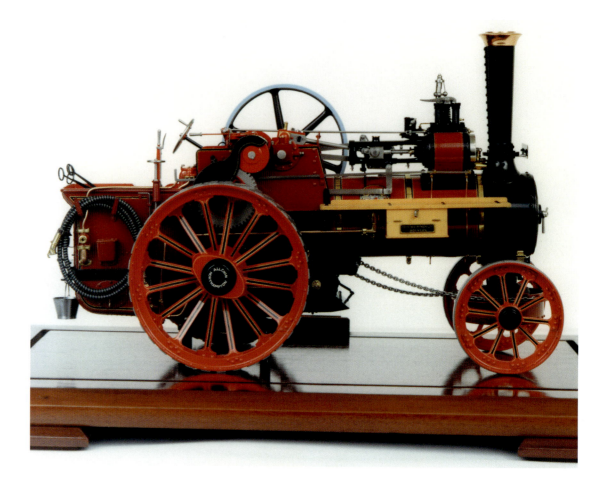

The Allchin Royal Chester traction engine – side elevation view.

At about this time Cherry started taking the *Model Engineer* magazine, just as Bill Hughes was starting his series on the Allchin traction engine, and bought a set of the drawings for 50 shillings (£2.50 for younger readers).

Cherry was not the only reader taken with the Allchin Royal Chester traction engine model at that time: many model engineers around the world built large numbers of models to the Bill Hughes design, making it the most popular traction engine model of all time. That original design is a fairly large engine in a scale of 1½in to 1ft or 1:8, which was a concern to Cherry. Her father suggested 'Make it half the size'.

So it was that Cherry's Allchin was built to a scale of ¾in to 1 ft, or 1:16. That choice was to play an important part later in her model-making adventures, becoming the scale of choice. Indeed it was used on every subsequent model apart from two where a larger scale was necessary to facilitate making parts which would not work in the preferred scale. It also made the Allchin project more difficult as there were no castings available in that size. The model is 12⅞ x 6 x 8in.

Royal Chester, No. 3251, is a single-cylinder general-purpose traction engine built in 1936. During construction of her model, Cherry tracked down the original engine, then at Leigh near Tonbridge. She was able to measure the important small parts on the original such as safety valves, lamps, drain cocks, cock handles, greasers, oil cups and fire irons.

The model occupied seven years of dedicated spare time up to 1964, when it was entered along with the Stuart No. 9 in the ME Exhibition. The Allchin took a Silver Medal, indicating that a talented model engineer had arrived. Two years later the model returned to the SMEE judges at the ME Exhibition where it was declared runner-up for the Duke of Edinburgh Award, the highest accolade in model engineering.

During 1966 Cherry completed a vertical twin steam engine based on the Stuart 10V. She had already started on a model of a Merryweather fire engine and wanted to be sure about the working of the reversing gear, so she built this little model which is three-eighths the size of the Stuart 10V.

Stuart 10V-based steam engine.

The Merryweather fire engine captured the imagination and marked the start of some serious model engineering to come.

EARLY SUCCESSES

Cherry made patterns for castings to be used in the model, and these were cast for her by apprentices at Stuart Turner, still in Henley-on-Thames at that time. Cherry recalls: 'Those castings were very good quality. No blow holes or any other problems.' A governor to a modified Stuart design was added to each half of the engine and a small 'Taper Lock' pulley to a Fenner patent fitted to the free end of the crankshaft. This little model was awarded a Bronze Medal in 1968.

Also completed in 1966 (having been started in 1964) was a model which captured the imagination of many when it appeared a couple of years later at the ME Exhibition at the Seymour Hall. This was a Merryweather self-propelled steam engine driven by a twin cylinder engine with twin pumps. This fine model also won a silver medal and the Bradbury Winter Cup, and again was runner-up for the Duke of Edinburgh Award. It was with this model and the Allchin that Cherry was pictured on the front cover of *Model Engineer* magazine, which was to change her life. That model marked the start of an extraordinary series of models which were completely original and led to her being recognized as one of the leading model engineers of the age.

Close attention to detail singled the model out and brought Cherry to public notice.

A rear view of the Merryweather fire engine.

CHAPTER FOUR

ON TO A NEW LEVEL

Aveling & Porter road roller 1931

Cherry Hill's first model to win a Championship Cup (forerunner of the Gold Medal) was of the Aveling & Porter Type AF 10-ton road roller No. 14277 of 1931. It also won the Championship Cup and the Crebbin Memorial Trophy in 1970. A year later it won the coveted Duke of Edinburgh Award.

Like most models from the 1970s on, it is to a scale of ¾in to 1ft, or 1:16.

The 'Type AF' denotes that this is a compound engine. The model was actually constructed as a twin-cylinder 'simple' engine; the low-pressure cylinder is fitted with a liner to reduce the diameter to the same size as the high-pressure one. Research on this engine began in the mid-1960s, based on information from works general arrangement drawings supplied by Mr E.A. Olive of Aveling Barford in 1965. However, all measurements and part

Controls and crankshaft. Note the nameplate.

CHERRY'S MODEL ENGINES

Aveling & Porter 10-ton roller.

details were taken from the original roller, No. 14277, owned by Mr Hervey-Bathurst of Eastnor Castle, near Ledbury in Herefordshire.

This project took model engineering on to a new level for Cherry Hill. The skills of manufacture are matched by the highest possible standard of finish. It is also a working model and has been run on compressed air to avoid heat and steam damage that would result from firing up the boiler.

This model is quite large even in the 1:16 scale, measuring 15 x 5½ x 7⅞in. It was built between 1966 and 1969, with some 2,500 hours being spent on the project over the period. Compared to some later models this one was completed fairly quickly.

An added point of interest is that it carries a family nameplate: George Hinds Ltd, Malvern, Worcestershire, was the company of Cherry's father and engineering mentor.

Rear view showing driving position.

CHERRY'S MODEL ENGINES

Aveling & Porter road roller model, which won Cherry's first Duke of Edinburgh award.

ON TO A NEW LEVEL

CHAPTER FIVE

AN INEVITABILITY

Burrell showman's engine 1922

Having built a traction engine and a steam roller there was a certain inevitability that the next engine would be a showman's engine. These are the Rolls-Royces of the traction engine world, the Blue Riband liners of the road, the Orient Express of the fairground. They were also true workhorses, generally running not just between fairground locations but twenty-four hours a day, seven days a week, driving the dynamo to provide power to light the site. It is not surprising then, that these were totally worn out at the end of their working lives. Given the nature of the engines, with their low, full-length roofs, twisted 'Olivers' and other brasswork and paintwork – a minor art form in its own right – it is not surprising that they have been much sought after by wealthy enthusiasts. In the 1950s, you could have bought one of these for £40: today you might pay 10,000 times that for a good one.

Cherry's model is of Scenic Showman's road locomotive No. 3909 built by Charles Burrell & Sons, Thetford, Norfolk, in 1922. Originally called *Prince Albert* the engine's name was later changed to *Winston Churchill*.

The design was based on general arrangement drawings published in *Steaming*, the magazine of the National Traction Engine Trust. All measurements were taken from the original engine, as was the paint scheme, which had been carefully researched by the owner when putting the engine back into original condition.

Cherry's favourite scale was not used for this model. Instead of 1:16 she chose 1:12 in order to permit making a working three-phase dynamo which would operate at the correct engine speed. That scale also permitted

AN INEVITABILITY

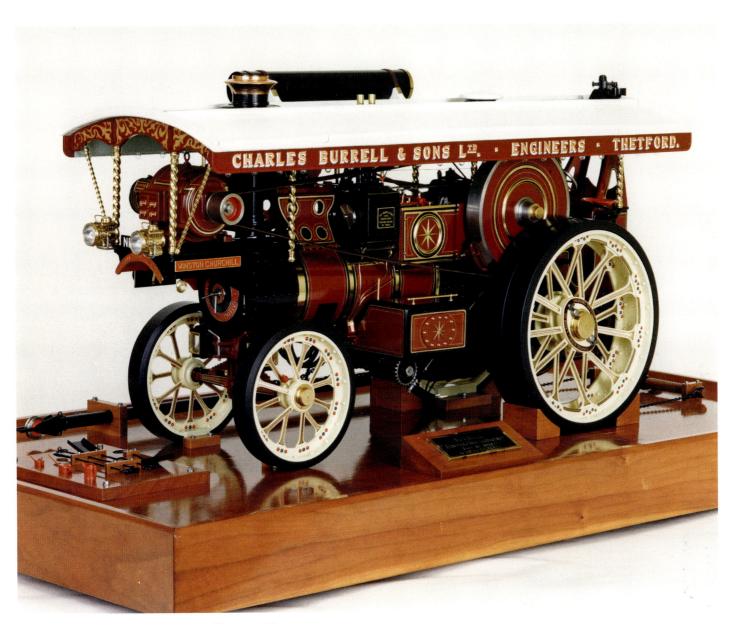

Burrell showman's engine – the magnificent workhorse.

Front view of the Burrell.

building an exciter (single phase) and a working, if not very accurate, pressure gauge. It further permitted some extraordinary detail, like the working padlock.

Cherry admits to not being comfortable working in this scale, and she returned to her favourite 1:16 scale for every subsequent engine, apart from the combined Savage Centre Engine and Organ Engine (see Chapter 9), which would have been too small to contemplate in her usual scale.

As usual the Burrell is a working model and the largest built, weighing some 66lb. It measures 24 x 9 x 12⅜in. It is coal fired and has run successfully on compressed air for extended periods. The model was built over a period of five and a half years from February 1970 to July 1975, occupying an estimated 8,000 hours gathering information, preparing drawings and in the workshop.

It is interesting to look at some of the detail of this model.

The generator is alternating current, based on a ceramic-type magnet and stator made and wound by Cherry using 34 swg wire with a diameter of 0.008in. When running at 750 rpm the generator lights up the front ten bulbs on the model, which are 1.5 volt, 0.11 watt.

The exciter is DC, based on permanent magnet shells. It incorporates a 5-pole armature wound by Cherry from 42 swg wire, with a diameter of just 0.004in. It lights up a 1.5 volt 0.11 watt bulb by the chimney at 1,350 rpm.

The voltmeter and ammeter on the model are dummies. The rest of the system is a working one. In total there are forty-two lights along the canopy, and they are also connected to a transformer hidden beneath the model's case. A switch under the front of the canopy will connect the front ten bulbs either into circuit with the transformer for static display, or into the model circuit powered by the generator.

The working coal-fired boiler has been pressure-tested to 90 psi, and has a working pressure of just 30 psi. Everything on the boiler works, apart from the injector, which proved too small internally to scale down to a working version, and so is a dummy. It has often been observed that 'nature cannot be scaled', although modellers like Cherry test the limits.

The drive and transmission of the engine is through a range of gears, all made by Cherry, including gears and drive pinions on shafts and the axle, bevel gears and pinions in the differential, and the steering worm and wheel. Also the bevel gears, worm and wheel on the governor. Unusually this governor does work, but not very efficiently at this scale. All of these parts involve some complex, advanced machining and the use of special cutters.

The engine is compounded and all passages are to scale. Effectiveness is governed by a 0.042in diameter passage through which the high pressure steam has to pass.

The model is fitted with a scale mechanical lubricator. This actually works when $\frac{1}{32}$in diameter wire which represents the pipework of the original is removed, and replaced with over-scale pipe with a 0.05in diameter bore. Low-viscosity oil is used.

Other parts which are dummies because they are not practical at the scale include the displacement lubricator, whistle and acetylene lamps, but they all receive full attention to

Supplying electricity to the site 24/7.

With the roof removed, it is possible to appreciate the work in the cylinders, drive and valve gear.

AN INEVITABILITY

Gears at the end of the crankshaft.

The tiny working governor.

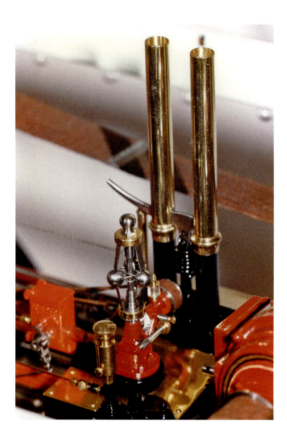

Fine finish on the driving wheels and detail of the lamps highlight the standard of work.

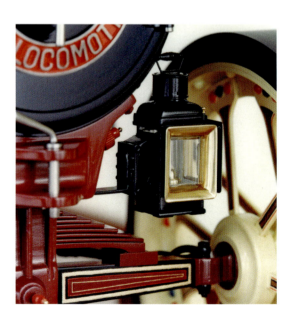

detail. The acetylene lamps, for example, each contain 105 individual parts.

The larger nameplates were made by cutting out the individual letters and soldering them to the back plate, and the smaller ones, the maker's nameplates on the valve covers and the instruction 'clean out boiler' were photo-etched by Cherry after learning the process.

Every last part on the model was made by Cherry.

AN INEVITABILITY

CHAPTER SIX

FROM MAKER'S WORKS DRAWINGS

Wallis & Steevens' Simplicity roller 1930

Cherry celebrated a return to her favoured scale of ¾in to 1ft with a pair of Wallis & Steevens' road rollers, the small single-cylinder Simplicity roller and its larger brother the Advance twin-cylinder engine. These delightful models occupied four years between 1975 and 1979.

The Simplicity was introduced in 1926 as a very light, 50 cwt roller for use on private drives, tennis courts, cricket pitches and so on. Despite its ability to work in confined spaces, it was not a commercial success against competition from small rollers powered by internal combustion engines. With its boiler set at an angle and clean lines compared with traditional rollers, the Simplicity reflected what was modern design in the 1920s.

Despite its name, the Simplicity incorporated a number of advanced features. At the time Cherry built this model, Wallis & Steevens was still in existence, simplifying the job of researching the engines. The general management of the company were most helpful during various visits to their works in Basingstoke, Hampshire, in the mid-1970s, and Mr T. Woods supplied various general arrangement and detail drawings. In addition, all measurements were taken from the original roller No. 8023 from 1930 owned by Mr R.H.U. Corbett of Hampshire.

The unmissable thing which made these engines stand out was the cylindrical boiler inclined at 27° to the horizontal. This ensured that the tube plate remained covered by water when on steep gradients which, if that

FROM MAKER'S WORKS DRAWINGS

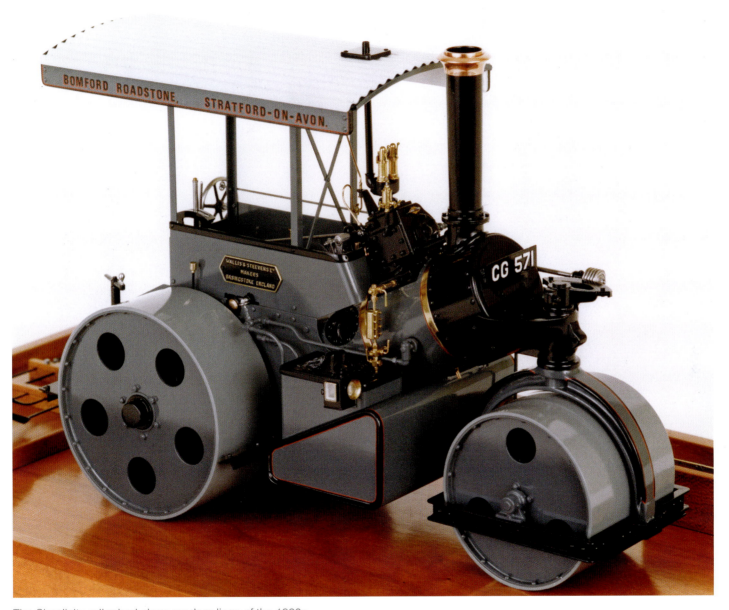

The Simplicity roller had clean modern lines of the 1920s.

The model includes all the mechanical features of the original.

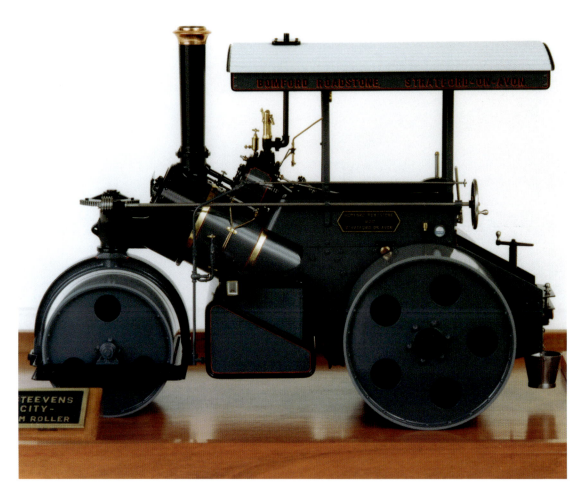

were not the case, might result in boilers exploding.

The Simplicity engine is a single cylinder fitted with a flywheel. It is a single-speed engine driving through a three-pinion differential on an intermediate shaft. Brakes are fitted inside the hind roll rims. The fore carriage of the Simplicity is of the overhead pivot type with overhead steerage. The model includes all these features.

The model has been run successfully on compressed air. If it is run on steam, water level in the boiler can be maintained by using a hand pump which has been fitted inside the right-hand water tank.

The size of the model is $9\tfrac{3}{8}$ x 4 x $6\tfrac{1}{4}$in.

CHAPTER SEVEN

FAITHFUL REPRODUCTION

Wallis & Steevens' Advance roller 1936

Wallis & Steevens' Advance road roller is a much larger engine than the Simplicity. As the name suggests it had a number of advanced features. The version modelled by Cherry is the 10-ton engine which faithfully reproduces all the engineering of the original.

The engine is a double high-pressure quick-reversing type with no flywheel fitted. It employs piston valves which are situated between the cylinders, has two-speed gearing and a four-pinion differential. Final drive is through steel spur gears direct to the rear rolls which revolve on their axle and each is positively driven. The rear axle features Advance Patent Automatic Cambering which permits rolls to adjust automatically to the contour of the surface being rolled. The rolls can also be set and held at a chosen inclination. The 10-ton hind rolls are cast iron, 4ft 6in diameter, ballasted with water; and the brakes are fitted to the insides of the rims.

The fore carriage is of the overhead pivot type with overhead steerage. This gives more positive control than traditional chain steerage.

Information for the model design came initially from a Wallis & Steevens Advance catalogue and from general arrangement and other drawings supplied by Mr T. Woods at Wallis & Steevens during factory visits in 1974–76. Subsequently all measurements were taken from the original roller No. 8100 courtesy of Mr T.C. Smith of Northants.

The model runs on solid fuel in the form of meta fuel tablets. Water is fed into the boiler via a hand pump fitted in the left-hand water tank.

The model measures $12\frac{1}{2}$ x $5\frac{1}{4}$ x $7\frac{5}{16}$ in.

Front view.

FAITHFUL REPRODUCTION

The model faithfully reproduces the original's engineering.

Double high-pressure engine with piston valves.

FAITHFUL REPRODUCTION

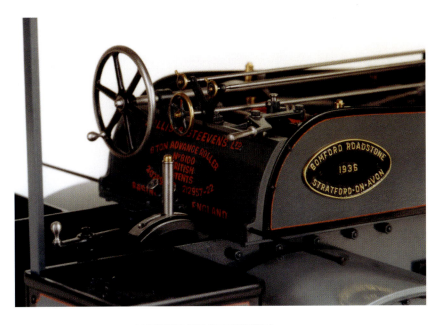

Above left and above
Overhead steerage replaces the usual chain arrangement.

Right side elevation.

CHAPTER EIGHT

A SEMINAL MOMENT

Taylor's Steam Elephant 1862

Starting work on Taylor's Steam Elephant in 1978 seems to have been a seminal moment in Cherry Hill's development as a model engineer. Hitherto, her models were based on established designs or on engines for which drawings were available and examples existed.

The Steam Elephant was a choice rooted in the study of Victorian documents, which also provided information, often incomplete, on the designs of traction engines from the nineteenth century.

From now on the critical time in the development of a Cherry Hill model would be the phase of research, delving into contemporaneous records to get close to the original, often by extending research into other developments of the same date to fill in details of what was missing in the available descriptions.

Work on the research into the Steam Elephant began while finishing the two Wallis & Steevens steamrollers. Research was followed by the design phase, producing drawings and testing them out on a mock-up model. (The one way to guarantee that everything will fit as it should, and come together as a truly representative model, is to build it.)

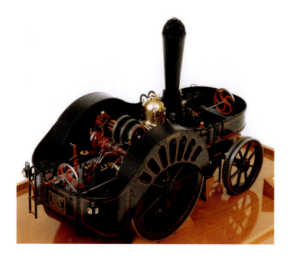

Right, rear quarter.

A mock-up of the Elephant was produced to a high standard before work started on the model proper. No parts used in a mock-up are transferred to a finished model, even though they are often made to the same high standards as the final version.

Builders of the original were James Taylor and Son of Britannia Works, Birkenhead, who manufactured steam winches, cranes and hoists before moving on to road locomotives. They took out patents for the Steam Elephant in 1858 and 1860 which were initially labelled 'traction engines', but later became 'road locomotives'. This was part of a movement in the evolution of transport described thus by *Artizan* magazine (1 July 1859): 'Steam engines intended to travel along highways for the purpose of drawing trains of loaded trucks or waggons, and also for hauling loads or agricultural implements over soft ground'. The magazine added:

> Now, Mr Taylor, who is well known for the admirable steam winches, cranes, hoists and such like labour-saving machines invented and constructed by him, has for some time past devoted himself to designing a portable steam engine, which shall be capable of running over ground of varying degrees of hardness, drawing loaded trucks or agricultural implements, and also of performing the duties of an ordinary portable steam engine for driving or rotating machinery, for raising or lowering heavy weights, and, by the application of a derrick or sheer leg, to perform the duties of a crane, besides having contained within itself the means of performing various other descriptions of work, as that of a crab, windlass, &c; it may be employed for pumping water or as a fire engine.

For all its versatility the engine is described as remarkably compact. It also had sprung suspension and a small turning circle.

A month later the magazine reported that the engine was in use at HM Dockyard

Right side elevation.

Right, front quarter.

Keyham. On trial it ran at 6 mph, light. At slow speed it hauled a load of timber which would previously have required twelve horses. It was also reported that a number of the engines were likely to be sent to India.

A year later the *Liverpool Mercury* reported that a Steam Elephant had been built for the Dutch Government for use in the docks at Flushing. This one was designed to draw a load of 40 tons on a level road and 15 tons up a 1:15 incline. This was not to the original patent of 1858; it included features of the second patent.

The second design to the patent of 1860 was shown at the 1862 Exhibition. It is not known how many were made.

James Taylor's first patent, 1858, was for 'Improvements in Portable Steam Engines'. The second, No. 1575 was for 'Improvements in Locomotive Engines and Wheel Carriages'. Both show vehicles of similar pattern with large (7ft 1in), diameter driving wheels and distinctive boilers, chimneys and spark arresters.

Cherry's model design was developed from the second patent drawings, the engine that was shown at the International of 1862, or the Great London Exposition, held, incidentally, on the site of what are now the Natural History and Science Museums. The building materials were reused at Alexandra Palace.

That patent showed an engine fitted with either three or four wheels. The four-wheel design was taken as the prototype for the model and the patent drawings formed the basis of the model design.

Cherry prepared a full set of working drawings from them; the general arrangement, various assemblies and details. These comprise twenty-two sheets sized 30 x 40in. It would be possible to make a full-size working Steam Elephant from these drawings. Cherry's model version was made to a scale of 1:16, or ¾in to 1ft.

Left, rear quarter.

Rear end steering wheel – another is fitted at the front of the engine.

It is interesting to note that the boiler is made as short as possible to control water level and the axle passes through the boiler, above the firebox. Thanks to the distinctive design of the boiler, water level was indicated by no fewer than seven level cocks.

Front and rear suspension was 'sprung', with the weight of the engine resting on India-rubber blocks.

As well as having a small turning circle, the engine could travel in either direction and was fitted with a steering wheel at each end. Normally the large wheels were leading to give easier passage over rough terrain, and the engine had two road speeds. It was fitted with gear drive to the winding drum, implying capstan-type use.

The chimney had a highly individual shape (the trunk, perhaps) and was fitted with a spark arrester, a highly public-spirited feature, given the high proportion of thatched and wooden houses still found around the English countryside.

Exhaust steam from the cylinders was fed into a tank which also contained the feed water. The patent claims resultant 'softened blasts' and reduced noise as well as providing heated feed water.

As work on the model progressed, its origins and shortcomings were revealed to Cherry. It seems likely that the design started with a steam winch, to which was added a 'suitably' shaped boiler, and a frame was then built around it. It was a remarkable achievement to build the Elephant in the workshops of that time. Although the design can be criticized today, Cherry has no doubt that Mr Taylor was a remarkable engineer.

However, it can be seen that he would have been beset with difficulties identified by Cherry, including:

1 Excessive load on the cross plates supporting the cylinders.
2 Slender valve mechanism – a characteristic of engines of the time.
3 Load on the chain drum when in use would distort alignment of the differential.

A SEMINAL MOMENT

It is likely that the design started with a steam winch.

Right Elephant front view.

Far right Rear view.

4 Drive to the main wheels via large diameter internal gears (popular at the time) would fill with stones and grit.

5 The cylinders and motion were in a difficult position for servicing, with the valve covers right up against the boiler.

6 Lack of space to clean out the smoke box and boiler tubes.

7 Passing the main axle through the boiler is not a good idea.

8 The length of the outside feed pipes from the superheater in the chimney base to the cylinders was excessive.

In spite of all the design shortcomings of the prototype the model runs well on compressed air. Work on the model occupied around 4,000 hours between September 1978 and August 1983 including painting, which involved many different shades of black. The model measures 13¼ x 7⅛ x 9¾in and weighs 32lb. It was awarded a Gold Medal and the Aveling Barford Trophy in 1984.

CHAPTER NINE

THE JOYRIDE

Savage fairground engines 1934

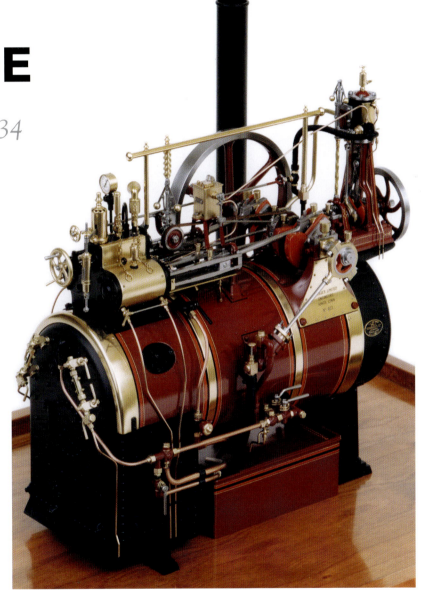

Savage's of King's Lynn is one of the iconic names in fairground rides, and this model comprises two of their engines: a No. 6 Centre Engine and a No. 4 Organ Engine, mounted at each end of a boiler. This pair of models were made to a slightly larger scale of 1:10 than the usual 1:16. The reason was that the Organ Engine is small and the valve gland studs were smaller than Cherry's limit for making such items of 0.6mm (0.024in) diameter, at which point the thread depth is only around three-thousandths of an inch.

This was the last centre engine made by Savage's in 1934. In this design it drove fairground amusement machinery via a belt on the flywheel fitted to the crankshaft, rather than an alternative which had bevel gearing driving a vertical shaft, which, in turn, drove through gearing the roundabout machinery mounted around the chimney.

Right, rear quarter.

Complete engine with Organ Engine at front end of boiler and Centre Engine at rear.

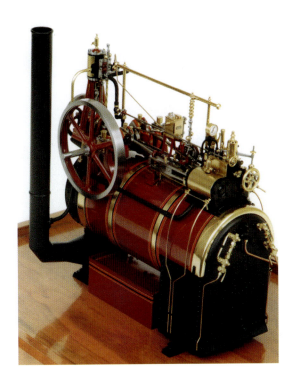

The second, smaller, vertical engine mounted above the boiler smoke box is a No. 4 Organ Engine which powered the music making.

Information for building the model was obtained direct from Savage's in the form of general arrangement drawings, with additional information from two authorities on these types of engine, Anthony Walshaw and Dr J. Middlemiss.

Measurements were also taken from the original engine, No. 903, in 1968 and 1981. Work on the model started in 1968, but was then left until 1982 and completed by 1984.

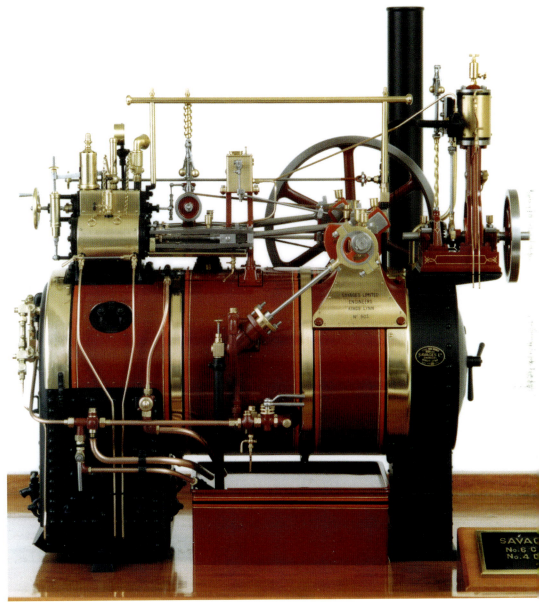

Right elevation.

The No. 4 Organ Engine.

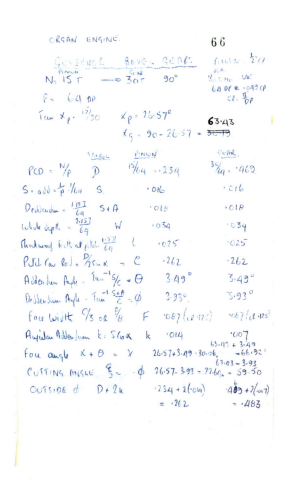
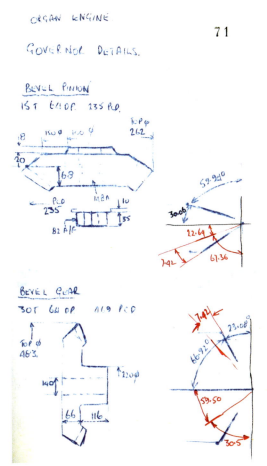

Overall it occupied some 2,500 hours of work. It won a Gold Medal in 1985.

Overall it measures 8⅜ x 5¼ x 8½in. The little Organ Engine measures just 2 x 1¼ x 2³⁄₁₆ in.

It is a working model, driven by compressed air.

Cherry makes use of duplicating notebooks to work out component details and keeps the carbon copies in the book to provide an aide-memoire when something similar is encountered in the design of a later model. Some sample pages are reproduced here from the Savage notebook to get a flavour of what is behind a Cherry Hill model. The pages cover some work on the governor for the little Organ Engine.

CHAPTER TEN

RETURN TO VICTORIANA

Batho 25-ton road roller 1870

For her next model Cherry selected another Victorian design, the Batho road roller. Research took place in 1981–82, while the Steam Elephant was being completed and before work continued on building the Savage Centre Engine. A further three and a half years of work between 1983 and 1986 to build the roller was rewarded with a Gold Medal and the Aveling Barford Trophy in 1987. It also won a Duke of Edinburgh Award in 1990. Cherry estimates that it took some 6,000 hours to complete.

The story of the actual roller is one of Victorian invention and enterprise which did not lead to fame and fortune. William F. Batho was a Birmingham consulting engineer who had designed a roller for the Calcutta city engineer which, after some difficulty in finding a manufacturer, started work in 1863. It may have been followed by two more machines for Bombay.

After that Batho joined Thomas Aveling. Aveling had introduced a new type of roller following trials in Hyde Park in London. It

RETURN TO VICTORIANA

Cylinders and valve gear.

Coal bunker behind front wheels.

had the rear steering rolls placed together to fill the gap between the two front rolls. The first of these new rollers was supplied to Liverpool, and Batho probably contributed significantly to the design.

Liverpool Roller No. 1 was a 12 nhp, 30-ton machine built to Patent No. 796 of 1867. It was a single-cylinder engine with the tank located over the rear rolls, and the steering wheel was across the line of the engine. It was reported that the steering was poor and the roller tended to crush the stone rather than compact it. That first roller was not a success and only one was built.

Subsequent rollers were modified. They were designed to accommodate two operators and another man had to walk in front under the provisions of the Highways Act of 1865. The steering position was moved to the left-hand side, just forward of the rear roll, and the whole steering assembly turned in line with the centre line of the machine. The tank was placed on the left-hand side next to the steering position. Brakes were fitted to the front right-hand roll and the weight of the machine was reduced.

This proved more successful and some seventy-five were made to this design, two of 25 tons, ten of 20 tons and sixty-three of 15 tons. Some thirty-nine were sold in the home market and thirty-seven were exported, to France, India and twenty-two to the United States. The last of these rollers was delivered

Drive from crankshaft.

Independent suspension and wheel spikes.

Steering on left-hand side and wood-block brakes.

Original Batho. The large roll is at the rear.

The model from a similar angle.

to the Duke of Portland in 1875. Thomas Aveling described the rollers in the Proceedings of the Institution of Mechanical Engineers in 1870.

Also in 1870 Batho redesigned the roller to incorporate two cylinders, front rolls mounted on separate shafts and driven independently through clutches, and spring loading on all rolls.

However, Aveling saw no need for the changes, and so Batho commissioned the roller to be manufactured by Thomas Astbury & Son, engineers and iron founders of Rolfe Street, Smethwick. Following Aveling's paper, Batho also described the roller in the Proceedings of the Institution of Mechanical Engineers. Thomas Astbury & Son is listed in the Birmingham Directory of 1872 as 'Smiths and Iron Founders of Ordinance'. Thomas lived in Handsworth and later the address listing was 'Mrs T. Astbury, Mr and Mrs Batho' – the only listing discovered for Batho.

There is just one photograph of the Batho, in poor condition. It was not a commercial success: only one was built. However, technically it was fascinating and an early attempt at a type of independent suspension.

Armed with a good deal of information from the Institution of Mechanical Engineers, Aveling Barford (Mr E.A. Olive), the single photograph, and listings from Birmingham Central Library, Cherry completed her research with a visit to the London Tram

Museum to take dimensions for parts of an Aveling & Porter tram engine of 1872 which contains some similar fittings to the Batho. Some differences were found between the drawings and the photograph, and the more suitable versions for the model were chosen. Originally, Weston's frictionless clutches were fitted but were not satisfactory in use and were replaced by 'ordinary solid clutches'. A working drawing of a clutch mechanism is included here.

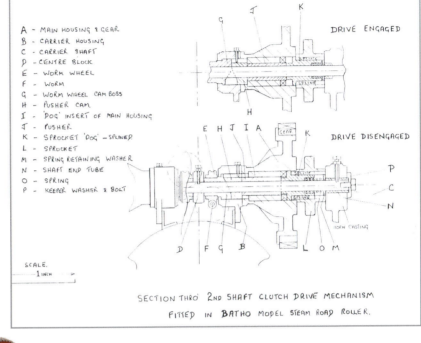

Cherry's drawing of Batho clutch drive mechanism as built in the model.

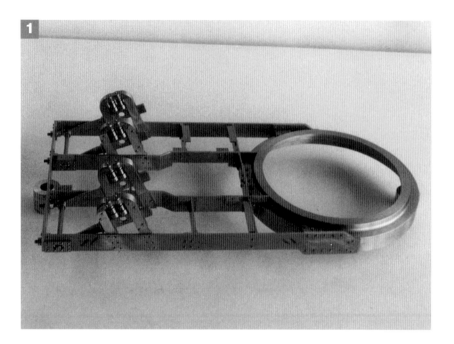

Batho under construction:

1 The frame and front axle support.
2 The water tank (left) and the coal box with tool box underneath.
3 Four-bearing crankshaft being lined up during construction.
4 Batho's elegant steering wheel.

The model runs on compressed air, is made to 1:16 scale, weighs 22lb and measures 15 x 7¾ x 9½in.

CHAPTER ELEVEN

A MODEL MADE TO WORK

Law & Downie road locomotive 1863

The fascination with the Batho was with its new ideas, even though they were not successful in the original prototype. The interest was the same with Cherry's next engine, the Law & Downie 1863 design, which was also highly innovative, but never actually built. In fact, it would never have worked because of some design shortcomings. Cherry, however, took up the challenge to produce it in a model form that would work.

John Downie, a mechanical engineer, had a successful start to a career which ended in tragedy. His story begins with the Crimean War (October 1853 to February 1856) between the Russian Empire and the Empires of the French and British, plus the Ottoman Empire and the Kingdom of Sardinia – the

The original Law & Downie engine could not have worked, but the model does.

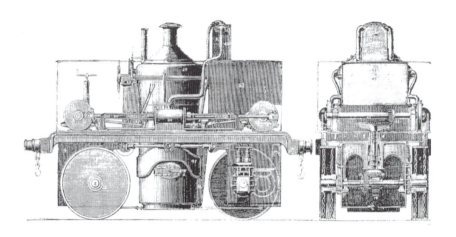

Drawing of the original from 1863.

Four-wheel drive and four-wheel steering, 1863.

first 'modern' war using the technology of railways and telegraph. It was also the first major conflict to be photographed.

Downie had a thriving business with government contracts to supply guns and shells during the war. But in 1857–8 his business collapsed and he took a job as general manager of the Phoenix Iron Works in Glasgow under owner David Law.

In 1863, Law and Downie patented a traction engine, but there is no evidence that the engine was ever made, and shortly afterwards Downie left the business when Law ran into financial difficulty. Downie then turned his attention to the manufacture of explosives. A meeting with Alfred Nobel, inventor of dynamite, led to the establishment of the industry in Scotland. Tragically he died, aged fifty-six, from injuries received in an accident at sea while destroying some damaged explosive.

Law & Downie's road locomotive represents another early effort at a technology that was to take many years to perfect: four-wheel drive. It also had steering on both axles. The model also has these features.

Mechanics Magazine of 2 October 1863 reported:

> In this improved traction engine there are two pairs of carrying or main driving wheels, and an important feature is the making of all four wheels both driving wheels and steering wheels. The steering is effected by inclining the axles horizontally in opposite directions, the axles being driven by gearing arranged so as not to be interfered with by this inclining of the axles. Each pair of wheels has mounted on it, on springs, a kind of bogie; and the two bogies support the main carriage framing,

A MODEL MADE TO WORK

Improved version for the model.

Completed model in unpainted state. Differentials at each end can clearly be seen.

Rear differential after painting.

and are connected thereto so as to be capable of swiveling [sic] for the steering action already referred to.

However, during construction of the model it became clear that some features needed to be changed if it was to work.

The patent showed that each cylinder had a single piston with extended rods, each end linked to the motion. As the crank pins of each crankshaft are set at 90°, it could not work. Correspondence to the *Mechanics Magazine* pointed out the problem. The puzzle for Cherry then was: 'How would Law & Downie have overcome this?'

The answer was to lengthen the cylinders, fit them with opposed pistons and introduce a shaft through the boiler with bevels to the vertical drive shafts at each end, to keep each end synchronized. That meant making new cylinders, valve chests and slide bars and extending the frame to suit. A result was that the brakes had to be redesigned as the bosses on the front slide bar supports were then where the handbrake pivot shafts were previously located. The model is fitted with brake blocks.

Other minor changes were made in the model. The steering as shown in the patent gave clockwise motion of the steering wheel turning the engine to the left. This was reversed in the model to the conventional direction. The patent also showed the engine completely encased in sheeting up to the

A MODEL MADE TO WORK

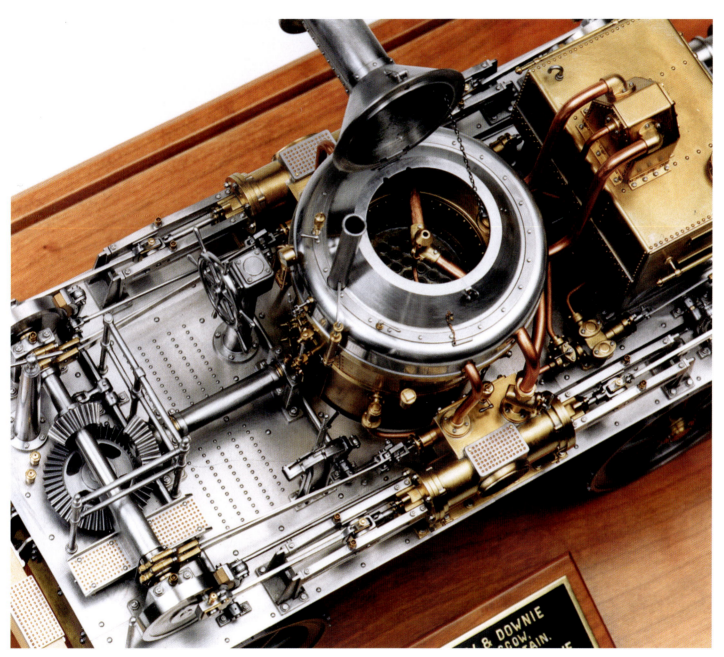

View of rear differential with the shaft running through the boiler.

CHERRY'S MODEL ENGINES

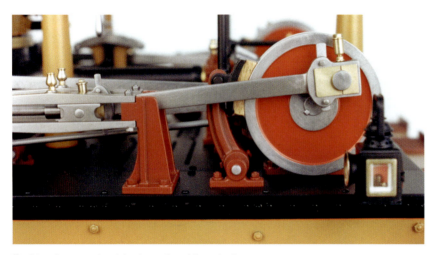
Braking by wooden block on the drive shaft.

Neat controls and authentic boiler fittings.

Right, front quarter.

smoke box level, with no method of access for the driver. On the model the sheeting is excluded and the driver does not have to climb over moving machinery to gain access to the driving position.

The model is to the preferred 1:16 scale, measures 14¼ x 6⅜ x 9⅜in and weighs 16lb. Research was carried out between 1984 and 1986, and design and build occupied three and a half years between 1986 and 1990. The model represents around 6,500 hours work. It was awarded a Gold Medal and the Bradbury Winter Memorial Trophy in 1992, and the Duke of Edinburgh Award a year later.

A MODEL MADE TO WORK

93

CHAPTER TWELVE

PARISIAN INCLINATIONS

Gellerat steamroller 1881

Research for this model of a *Rouleau Compresseur à Vapeur* by E. Gellerat & Cie, Paris, was extensive and taken from a variety of sources, including:

- French Section of the Foreign Patent Office (London Reference Library)
- *Institut National de la Propriété* (French Patent Office) Paris
- Institution of Mechanical Engineers
- Science Museum Library
- British Library
- French National Railway Museum, Mulhouse.

Eugène Gellerat developed his steam roller from patents of M. Balaison of Bordeaux dated 1860. It had a locomotive-type boiler which provided power for two one-piece rolls, each mounted on an axle. The axles were unusual. One end was spherical and fitted into a matching housing and the other end was held in a suspension unit which could be moved along the chassis. The rolls were arranged so that they could be inclined out of parallel with each other to steer the machine.

PARISIAN INCLINATIONS

Steering wheel.

CHERRY'S MODEL ENGINES

Gellerat cylinders inclined at 45°.

Rollers are driven by chain and sprocket.

M. Gellerat took out five patents for complete steamroller designs and one for a *Rouleau Compresseur Electrique* in 1883, plus other patents on improvements to other designs.

Cherry's research for the model took place between 1986 and 1989, and construction was 1988–91, although painting was completed only in 1995.

The model is based largely on the drawings for the patent of 1881 with the roller's cylinders mounted on each side of the boiler and inclined at 45°. Roller drive is via chain and large sprocket. The roller axles do not turn and one end, which is spherical, is located in a special housing to which the sprocket is attached. In the roll and parallel to the axle is a shaft called the 'toc' that projects into a cushioned bearing to take up displacement as the roller moves.

A number of parts are not shown on the 1881 patent drawings. The steam fittings and injectors were based on an 1867 print of an earlier Gellerat machine in *The Engineer* (these are non-working dummies on the model). That was also the source for the design of the model's chimney, chimney closing plate, tank filler, and lamps. The safety valve is from an 1873 patent.

A number of items on the model were designed on the lines of those found on French railway locomotives of the period. These include the smoke box doors and latch, chimney cap, spark arrester, steam dome, regulator and some of the steam fittings.

The entire model is made from metal stock. Complex shapes are fabricated; no castings are used. Every last part was made from scratch by Cherry including all the transmission gearing, steering bevel gears (15- and 27-tooth 60 dp), nameplates, all visible nuts, bolts and studs, lamps and the small chains on the tool box wedges made from soldered 0.008in wire. The rolls were finished using a home-made sand-blasting setup. The boiler is made in brass. M. Gellerat wrote that his boilers were fitted with copper fireboxes and brass tubes.

The model was awarded a Gold Medal and the Bradbury Winter Memorial Trophy in 1998 and the Duke of Edinburgh Award a year later.

Steering operated by bevel gears.

Gellerat engine right-hand side.

Gellerat left-hand cylinder.

CHERRY'S MODEL ENGINES

CHAPTER THIRTEEN

THE GREAT PROJECT

*Andrew Barclay's traction engine
and boring engine 1862–3*

The combined units of the Andrew Barclay Traction and Boring Engine represent the largest project undertaken by Cherry Hill. Research began in 1988, and design and construction occupied some 9,000 hours of work between 1990 and 1994. They were shown in 1998, winning a Gold Medal and the Bradbury Winter Memorial Trophy and, the following year, another Duke of Edinburgh Award.

Andrew Barclay was the son of a millwright. He was apprenticed as a plumber and coppersmith, but was more interested in engineering at his father's works. In 1840 he set up an engineering works in Kilmarnock with a partner, but started his own business next door two years later. Barclay patented and produced a wide variety of steam products including high- and low-pressure engines, colliery winding and pumping engines, cranes and pile-driving machinery. The business grew and prospered. The first railway locomotive was produced in 1858 and 300 were made over the next twenty years.

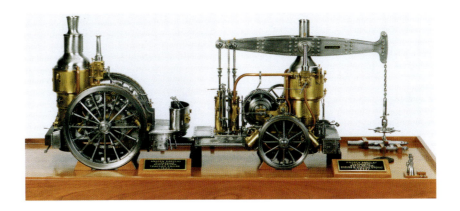

The engine before painting.

The traction engine is on the left and the boring and winding engine on the right.

Throughout his life, Barclay was interested in astronomy and produced a number of telescopes. He was made a Member of the Astronomical Society. Unfortunately this fascination with telescopes caused the engineering business to suffer and he ran into financial difficulties in 1874, and was further affected by a serious fire two years later. Eventually the company was reorganized under new management and a limited liability company was formed in 1892, ending Barclay's control of the business. He died eight years later at the age of eighty-six. The company continued to build locomotives, latterly as Hunslett-Barclay and Brush-Barclay. Today the Kilmarnock Works provides rail engineering services as Wabtec Rail Scotland.

Prior to his difficulties, Barclay patented two traction engines in 1862 and a boring and winding engine in 1863. These patents provided the basis for Cherry's models.

Research took Cherry to a number of sources including the Mitchell Library, Glasgow; Glasgow University archives; the Dick Institute, Kilmarnock; the London Reference Library; the Institution of Mechanical Engineers; the Kilmarnock works; and the late Tom Walshaw (the model engineer who wrote under the pen name Tubal Cain). The patents and further reading provided more information.

Surprisingly, perhaps, there was no record of traction or boring engines among the

The traction engine.

The boring and winding engine.

Inner gear ring and two-speed gears visible through driving wheel of the traction engine.

extensive Barclay records at the Dick Institute, and there was no record at the Andrew Barclay Caledonian Works in Kilmarnock. It is also not certain whether any traction engines were actually built, although it is known that at least one boring and winding machine was.

Traction engine patent 646 of 1862 describes two designs designed to haul industrial loads on ordinary roads. One of the designs is for a three- or four-wheeled traction engine with a locomotive-type boiler, two large diameter driving wheels, and one or two steering wheels to the rear. However, it is the other two-wheeled engine that was chosen for the model.

Each wheel is driven independently by a crankshaft across the frame from two cylinders arranged in inverted V-form on either side of the engine. The wheel side of the shaft is fitted with spur gears. These engage, via radially mounted idlers, with either the internal or external toothed gears attached to the wheels to give a choice of high or low gear.

This method of gear engagement is Cherry's interpretation from Mr Barclay's description that either gear in each wheel may be engaged by 'lifting or moving' the crankshafts. This presented problems with valve travel for the model. Total distance to lift the full-size crankshaft is 2¾in and 0.168in for

The regulator on the side of the boiler, seen here before painting, can be used for steering.

the model. The model is in Cherry's usual scale of 1:16.

Three alternative methods of steering were described. The primary method was by regulating the steam supply to the pairs of cylinders on each side of the engine. The regulator is fitted with a cylindrical valve. As well as the standard in and out movement to admit and control steam into the cylinders, the lever can be moved left or right to increase steam flow to one side or the other for steering. This proved to be a most satisfactory and flexible arrangement in tests on the model.

Alternative steering methods are a) altering the positions of reversing levers controlling either pair of cylinders, or b) applying brakes to either crankshaft. The boiler is a submerged top tube plate type with a superheater. The patent drawing did not include a firebox door. The model has two: one on the front of the boiler for ground-level firing and an extension to the firebox under the crankshafts for use by the fireman while travelling, which required the original brake design to be modified.

Other items on the model not shown on

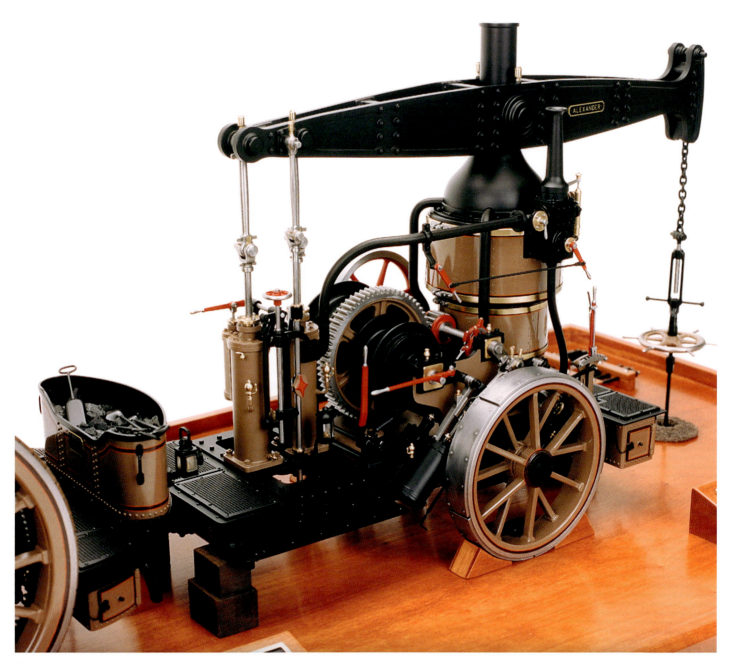

Boring and winding engine.

the patent are: front coal and jack stowage boxes, ladders, tool boxes, and cosmetic fittings.

To ensure the model was a working one it was necessary to provide satisfactory clearances and thus to alter the position and action of the reversing levers. Only one was shown on the patent, but if the second were drawn it would obstruct the rear crankshaft. Modifications were also necessary to give clearances for the cylinders, valve chests, big ends and eccentrics. Also, the ratio of the connecting rod:stroke was increased from 1.6:1 to 2:1 to bring it more into line with traction engine practice of 2.5:1 minimum.

All parts of this model, and the winding engine, are made from solid stock metal or are fabricated. No castings are used. Every last visible part is made by Cherry, including the small chains for the tool box wedges and gear change pins. Each link is soldered. Links are 0.048in long and made from 0.008in diameter wire. Other parts made by Cherry include nuts, bolts and studs. Also lamps, nameplates, chequer plates and gearing.

The boring and winding engine is designed to be trailered by the Barclay traction engine described above, and is one of two described at the time. The other is a self-propelled four-wheel version.

A Barclay Patent Locomotive Mineral Boring Engine is described in the *Colliery Guardian* of 6 December 1862. Andrew Barclay states there that 'Having purchased Mr Simpson's Patent Right, which is now combined with my own, the public is respectfully informed that I am now the only maker of these Direct Acting Beam Boring Machines.'

The article went on to state that the machine is capable of boring to a depth of 300 fathoms (1,800ft or 549m) and that it can work at any length of stroke between 1in and 2ft with 'any weight of blow'. It also included letters from satisfied customers.

George Simpson described the design of his machine in a paper to the Institution of Engineers of Scotland. Operation employed the use of shear legs and pulley to lift the boring rods which were 27ft, 18ft and shorter. He also described types of boring tool and pump for collecting debris. This long paper further describes methods for collecting samples of each stratum of rock to be bored through, the constitution of seams, depths of deposits and quantities of coal, ironstone and other mineral deposits.

Andrew Barclay's patents for boring and winding machines describe two engines. The first is a self-propelled engine with a locomotive-type boiler. The second engine has a vertical boiler, powering a beam of 15ft supported through its pivot axis on the chimney. Drawings for this second engine show details of the trailed engine chosen for Cherry's model.

Boring cylinders with regulator on the left and winding drum brake on the right.

Far right
The boring cylinders before painting.

In these drawings the boring cylinders (actuating and balance) which operate the beam are mounted on a platform at the front. The drawings show the valve gear for regulating the motion of the actuating cylinder and adjustable tappet positioning to give variable stroke for the actuating cylinder achieved via a vertical threaded shaft and hand wheel.

The winding engine is similar to the traction engine, with inverted V-cylinders which are mounted on the left-hand side. Two winding drums of different diameters are mounted between the frames, driven from the engine crankshaft. Each drum has its own clutch and brake. To use the drum to withdraw a boring tool, the cylinder and valve actuating rods are disconnected and the beam swung round. The operation requires use of a 'pit head frame' with suitable pulleys and chains.

Patent drawings show two types of beam, one of plate construction and one cast. Two trial beams were made and it was decided to

THE GREAT PROJECT

Inverted V-cylinders.

use the cast version (but fabricated for the model). On the original the balance cylinder was open to atmosphere at the lower end. When operating the model, balancing was only possible with the bottom of the cylinder closed with a cover which was fitted with a small bleeder.

The drawings give no water tank details or method of feeding water to the boiler. An injector similar to the one on the Barclay traction engine was fitted. As mentioned previously, other items not shown on the drawings include: boiler fittings, coal box, tool box, side platforms, steps and various small fittings. Research into contemporaneous items provided the basis for the designs used on the model.

Research and initial design of the boring and winding engine was carried out in 1990 and 1991. Final design, drawings and construction took place between 1991 and 1994. The model was successfully tested running on compressed air. It was painted in 1998.

An estimated minimum of 4,000 parts are incorporated. All parts are made from metal stock or fabricated; no castings were used. As with the traction engine all small parts including nuts, bolts and studs were made from scratch, and the small chains were constructed with silver-soldered links. The model traction engine and the boring engine each weigh 10lb and together occupy a case measuring 29 x 14 x 13⅛in.

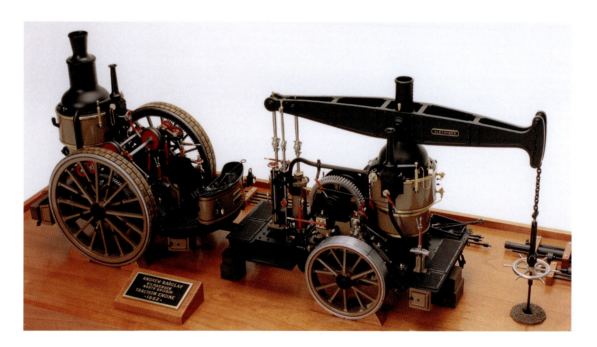

CHAPTER FOURTEEN

A FAMILY AFFAIR

Gilletts and Allatt traction engine 1862

Model engineering taken to the next level.

You will have seen earlier that Cherry's father and mentor was born at Gilletts Farm in Smarden, Kent. You will also have seen that her mother's maiden name was Allatt. That gives a clue to the inspiration for this model, which takes model engineering to the next level.

Cherry's early models were made from commercial drawings, then designs were prepared from prototypes that still existed and could be seen and measured to ensure model accuracy. Next, the subjects for models were only available from contemporaneous published articles and patents, often with incomplete information. The Gilletts and Allatt traction engine, however, is an original.

Cherry wanted to design and patent a traction engine as it might have been in 1862. The idea was to incorporate features not taken up by other designers at the time. That is how Patent No. 2260 (a number from the time but not allocated then) came about.

George Gilletts was born in 1825, the youngest of six children. His father, William, owned an estate of 190 acres which included four farms and the village forge. One can conjecture that young George spent many hours with the blacksmith, and naturally turned to steam and traction engines to benefit the estate. His patent, jointly held with Henry Allatt, mentions a further patent dealing with 'Improvements to Traction Engines'.

George Gilletts' and Henry Allatt's Patent No. 2260 describes a traction engine comprising two separate frames fixed rigidly together to form one four-wheeled engine. The arrangement chosen for the model was that with two identical frames, thereby forming a traction engine with four-wheel drive and two-axle steering. The front frame, which contains the boiler with the cylinder attached, water tank, transmission gear and steering arrangement, is described as the 'Driving Frame'; the rear frame is referred to as the 'Loading Frame'.

The patent states:

> By reason of the simple method of fixing the two frames together (the insertion of three or more locking pins of substantial diameter situated in line across the main axis of the Engine), it follows that the design of the Loading Frame can be that most suitable for the work intended. Hence the operator has a choice of different Loading Frames. Furthermore time is saved. On reaching the destination with full load, the Driving Frame is separated from its load and can be immediately attached to a previously loaded Frame for the return or onward journey. In cases when an Engine of superior power is required, two identical Driving

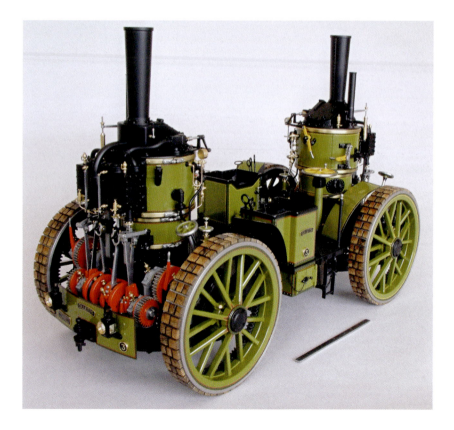

Four-wheeled drive for the Gilletts and Allatt.

Engine, front view.

Frames are locked together. Thus the Engine becomes a four-wheeled Traction Engine with each wheel driven and each axle supplied with arrangement for steering. The load is towed in the manner of Traction Engines.

The Driving Frame consists of an operator's platform, and at one end is attached a large diameter wrought-iron ring of right-angled section. This ring is supported by means of small vertical and horizontal bearing rollers attached to an inverted inner angle ring in the same horizontal plane. The inner ring supports the axle and suspension, double-crankshaft, gearing, and secondary shafts driving each wheel. Either wheel can be made free of its drive by the disengagement of gears on the shafts. Centrally supported in the inner ring is a vertical multi-tube boiler, on the forward side of which are vertically mounted the cylinders, beneath which are the supports and rods necessary to give motion to the crankshaft. The engine has the usual eccentrics for reversing the motion. Drive to each wheel is by engaging a gear wheel fitted at the end of the intermediate secondary shafts with an annular-toothed gear wheel secured to the inside of the wheel.

The patent continues:

> In the peripheries of the driving wheels are cast rectangular or square recesses into which are driven end-grain wooden blocks. The blocks when fitted protrude from the surface of the wheel; the recesses may be cast in line or alternatively pitched, and varying in number across the face of the wheel.
>
> Centrally on the rear side of the inner angle ring is a worm-toothed segment, which engages with a worm wheel attached to one end of a horizontal shaft mounted on the outer ring. On the other end of this shaft is a small bevel pinion which engages with a similar pinion fixed at the end of a vertical steering column. Brakes may be fitted independently to

Driving wheel with ring gear inside. A brake band operates on the drum on the axle. The secondary shaft and crankshaft can be seen on the right.

Cylinders are mounted vertically.

each wheel. The action is obtained by the contraction of a band around a metal drum securely attached to the wheel.

The chimney is set upon the smoke box forward of the centre-line of the boiler to allow for easy cleaning of the boiler tubes. In the smoke box ends the exhaust pipe from the cylinders which acts to create extra draught. The draught can be further increased by the introduction of a blower. Water is fed into the boiler by means of an injector.

CHERRY'S MODEL ENGINES

End view of the unpainted model showing clearly the cylinders and crankshaft.

A FAMILY AFFAIR

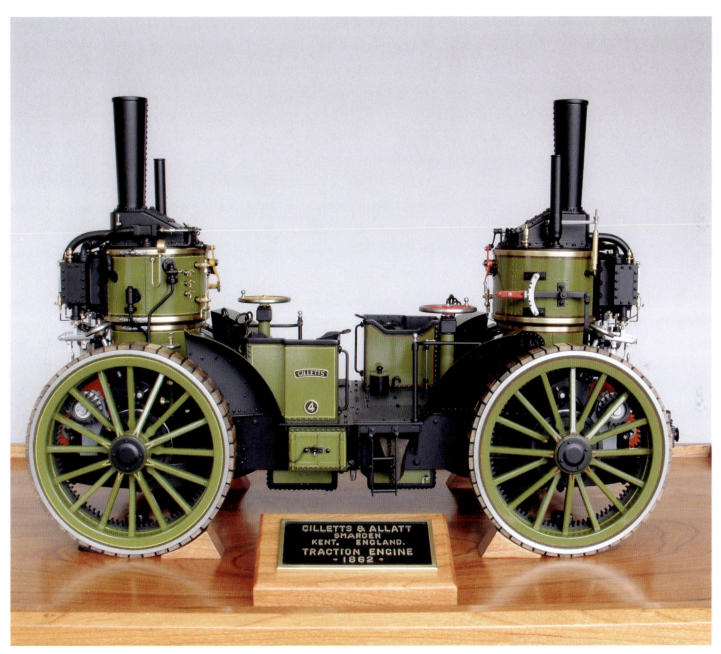

Side elevation.

Driver's platform with steering wheels for each axle.

The boiler has a totally submerged top tube plate, and the water tank capacity is equivalent to 170 gal. Coal carrying capacity is 7 cwt.

Research and initial sketches were done during 1990 and 1993. The final design, drawings and construction, involving more than 7,800 individual components, took five years between 1994 and 1998 with another five and a half months to paint the model in 2001.

Some items are not shown on the patent, including the boiler fittings, the position of gear engagement and the injector position.

Operator's items not shown include the floor plate access, steering lock pins, tool box, wheel guards, and motion shield.

As with all Cherry's later models, all parts were made by her, including gears, chequer plates, nameplates and every last nut, bolt, stud and chain link.

Having designed the original, Cherry made the model, which was shown in 2002, winning a Gold Medal and the Bradbury Winter Memorial Trophy. A year later it won the Duke of Edinburgh Award.

CHAPTER FIFTEEN

THE ENGINE IN A WHEEL

Blackburn agricultural engine 1857

The 1857 agricultural engine of Isaac and Robert Blackburn was a 'natural' for Cherry Hill. To say that it is unusual is an understatement. It had one large driven wheel with the boiler and cylinders inside. To produce a model of one of these extraordinary engines would be tricky enough with detailed drawings to work from. Unfortunately, the only available drawings were some way short of explaining how the thing worked – if it actually worked, that is. The model certainly does work.

Why build a model of an unsuccessful piece of engineering? It's the same motivation that makes people climb mountains or restore steam locomotives – because they can. In the case of the Blackburn, because it represents a flight of engineering fancy from an era when every experiment had to be conducted, every

Boiler, cylinders, almost everything is inside the single driving wheel or drum.

Blackburn 1857 engine – a Victorian flight of fancy.

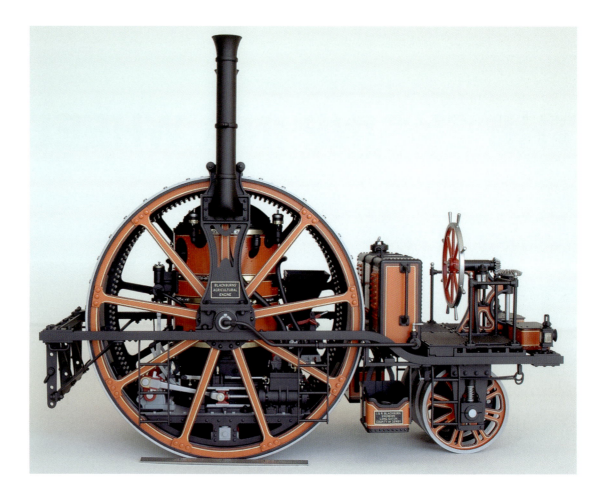

idea tried, every creative thought explored. And there is the added challenge of making it work.

Here in the twenty-first century it is easy to look at the design and predict that it would not be a success. But in the 1850s steam engineering was still in its infancy. Criticizing it today it is like the owner of the latest high-tech automobile being critical of a model from the 1890s. In their time, all ideas like the Blackburn were exciting, and people of the day would have found it difficult to guess their impact on the future.

Inspiration for this model came from a favourite source, the *Mechanics Magazine*. Made in Cherry's usual scale of ¾in to 1ft (1:16), the Blackburn is far from simple. Traction is applied through one great wheel, with

THE ENGINE IN A WHEEL

the power applied from an engine located inside that wheel – and the boiler is inside the wheel, too. The only element of the power plant of the engine outside of the wheel is the pair of elegant chimneys, one on either side. How they work, and some of the other tricky arrangements on this complex engine, were puzzles to be solved.

Research was done through Patent Office records and contemporary publications, such as *The Engineer* and the lesser-known *Mechanics Magazine*, which is a real treasure trove for research into Victorian machinery. For some models there is a further stage of finding some works drawings, or even an example of the engine to be modelled, or the remains of one that can be examined and measured. As with other models the research for this one went into items like contemporary valves that might have been fitted. Also, into other products of the maker of the original, to give yet more clues to producing something as authentic as possible.

So, what did the research into the Blackburn come up with? Certainly not an actual example, as it is quite possible that none were actually made. *The Engineer*, in September 1857, showed an engraving of the first engine and gave a short description. That engine was the subject of Patent No. 414, granted in the same year to Isaac Blackburn of Islington in the County of Middlesex, and Robert Blackburn of the City of Edinburgh, for

Front view of steering and of chimneys, which are not connected to the smoke box.

A view through the drum with the axle passing through the boiler and the cylinders below. Drive is via the large ring gear inside the drum.

THE ENGINE IN A WHEEL

'Improvements in Engines or Implements to be employed in Agriculture, applicable also to the Transportation of Heavy Bodies, to the Traction of Carriages and the Conveyance of Passengers.' These contemporary sources provided the following description of the main elements of the Blackburn engine, and helped to unravel the mysteries of how it worked, or perhaps that should be 'could have worked'.

A vertical boiler and a pair of cylinders are placed inside a drum of 'considerable diameter', the drum being driven by an 'internal gear or other gearing'. 'Friction wheels' are fitted below the engine and the boiler plate, which run on rails inside the drum to 'give steadiness to the drum'. That implies that they took part of the weight of the machinery inside the drum and prevented distortion. The smoke box extends each side to just inside the drum spokes. Immediately outside the spokes are the base brackets of the chimneys.

How did the smoke get from one to the other? Replying to correspondence in *Mechanics Magazine*, Mr Blackburn stated that there was a narrow gap between the openings to allow the spokes to pass through, and that the smoke 'readily finds its way therein and of course is carried upwards'.

The regulator which adjusts the speed of the engine is 'situated in the boiler' and is operated through the trunnion on one side of the drum. The other trunnion is employed to feed water to the boiler from the tank on the

The boiler, cylinders, drive and coal hopper removed from the drum.

121

mainframe in front of the drum. Inside the drum is a 'Fire Feeding Apparatus with Coal Hopper'. The drive pinion is shown but there is no written information in the patent. The boiler has conventional fire doors. The patent also states that the engine could be used as a stationary engine by disengaging the toothed gear – however, it does not explain how. The patent also covered reversing the engine through a drum trunnion (although the drawing shows a separate lever for that, inside the drum). It also included an adaptation of the engine to run on railway lines.

The Blackburn was by no means the first 'moving drum' design. *Mechanics Magazine* reported in 1857 that there had been attempts to produce one over a period of years. Did it ever appear? There is an unconfirmed report that a Blackburn engine was shown at Salisbury in 1857. However, it failed to make promised appearances at the Royal Agricultural Show at Smithfield in the following two years.

Cherry analysed some of the reasons for the failure of this engine. The initial assembly would have been difficult; servicing and setting had to be performed through the spokes of the drum, and the frame section is too light and would have flexed significantly during operation. Recognizing that, the model was built with a larger than scale frame, but still flexes.

The pump and the automatic stoker could not operate while the engine was stationary, and firing through fire doors on either side of the boiler would have been difficult, because of their location. The machine had to be stopped to adjust boiler feed and safety valves, and to shut off the drain cocks. Moving the reversing lever or disengaging the stoker meant reaching inside the drum which, of course, would not be possible while the engine was moving.

Undeterred by all the difficulties shown up during the research, Cherry proceeded to the design stages. Research and initial sketches for the model were done between 1996 and 1998, while also finishing off a previous model. The design, drawing and construction occupied the following six years, including making a mock-up of the engine, with many parts fully machined as on the model proper. Painting took a further four and a half months, with completion just in time for the 2005 Model Engineer Exhibition. Something over 5,200 components went to make up the Blackburn.

For Cherry it is this research and design phase that is the most fascinating part of tackling a new model. The actual construction is less so, even though anyone looking at her models will have their breath taken away by the craftswomanship, the meticulous care that goes into the research being matched by the care in construction. For example, all the rivets are hand made, all the 'castings' are actually fabricated, and boilers are made of

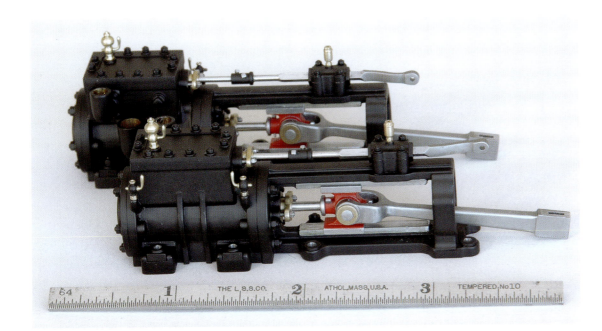

The pair of cylinders used on the model.

steel rather than an easier-to-work but non-prototypical metal like copper.

Anything less than perfect is rejected. Cherry has a Blackburn cylinder that she rejected after it had been completed following goodness knows how many hours of work. Handing it over, she says: 'You'll see what's wrong with it.' Examining it closely and searching for where the end mill dug in or the drill broke through or, well, the sorts of things most model engineers do, nothing is apparent. After admitting defeat, she points to one of the bosses in the cylinder 'casting' into which the cylinder cover bolts are screwed. Still baffled. Eventually she explains that it is about a millimetre too long. So it was scrapped.

Equally disarming is the response to questions about how this or that was made: 'Oh, I just fiddle around until I get it right.' That doesn't mean that anything is casual. During construction she keeps notes of design details and how many of the parts are actually made. Designs are carefully and comprehensively drawn to a standard you would expect from a draughtswoman who had twenty years experience with an agricultural machinery company.

With the Blackburn, the design process started with a general arrangement drawing of the original. As building progressed a number of modifications were necessary, not least to produce a working model. Many people looking at the Blackburn and the other traction engines may think they are produced just to look right. Not so. They have to be able to work, and they are tested.

As work progressed on the engine a number of modifications were needed to produce the model. The following details give an idea of the challenge inherent in modelling in great detail, even from a works drawing. The diameter of the lower section of the boiler was reduced to clear the trunk guide housing, to fit the fire door with clearance for firing, and to leave enough space for the movement of the expansion link and primary valve link. The regulator is mounted externally on the boiler, and is operated through a drum trunnion. Additional linkage was given to provide a second operating position.

You will never see it, but a shield is fitted inside the boiler by the regulator to ensure that only steam enters the regulator. The same applies to the safety valve outlets – two safety valves having bronze seatings were fitted to the model. It was necessary to alter the position of the weighshaft and reversing lever to give sufficient clearance for operation. A quadrant assembly was fitted.

The ash pan was made so that when dropped there is the equivalent of a 3in gap for emptying. Emptying would have been difficult, even so.

On the patent drawing the large internal drum gear is shown in eight sections. The one for the model was made in one piece, with pads where the break points would have been. Construction time just for this one gear and all the tooling necessary took some two and a half months.

Inevitably the business of stoking a boiler situated inside a large wheel, without stopping the wheel, requires some ingenuity. There was no information for the final drive of the stoker and its internal parts. The patent stated: 'The Boiler is furnished with a fire-feeding apparatus with a large hopper to hold fuel, which is set in motion by gearing of the drum, thus enabling the engines to work for two or three consecutive hours without stopping to fuel the furnace.' The model stoker is fascinating in its own right and was interpreted from the patent information. It has a pair of rolls with spikes interspaced and set at 45° to each other. The outer roll is spring-loaded. It is operated by engaging a ratchet pawl and opening up the flap door on the boiler. The foundation of the stoker assembly is a support frame attached to the engine base plate and the boiler.

Two side plates house a pair of feeder shafts, one of which is spring-loaded. The feeder shafts are fitted with the rods that feed the coal and restrict the flow from the hopper. Feeder shaft gears are 17-tooth 48 dp. The two side plates are spaced just 0.64in apart. The main drive to the feeder is an auxiliary shaft with a 15-tooth 24 dp gear, and is driven from the large internal gear in the drum. There is an eccentric boss on the auxiliary shaft with a drive rod to the drive plates of a 32-tooth ratchet wheel on one feeder shaft. The pawl arm can be disengaged.

The chute from the feeder to the firebox has a vertically suspended baffle gate at the delivery end. While feeding, the gate is pivoted outwards and secured. When feeding is complete the gate is lowered to cover the hole in the firebox, and that also clears any coal left in the chute. During testing the chute remained clear at all times.

The feed rate can be adjusted by removing selected rods from the feeder shafts. The model was tested with maximum feed.

The hopper volume is just under 1in^3. and that is just over ½oz of coal (equivalent to 1¼ cwt in full size). Its sides are steeper than those shown in the patent drawing. Coal used to test the stoker was West Virginia coal, riddled through a grid 0.061 x 0.165in. When tested, it took 373 turns of the stoker main drive shaft to empty the hopper. On completion the model stoker fed the boiler satisfactorily, at an acceptable rate, and there was no 'bridging' in the hopper.

It was no surprise that the 1857 Blackburn model was awarded a Gold Medal and the Bradbury Winter Memorial Trophy in 2005, and the Duke of Edinburgh Award in 2007.

Cherry kept a photographic record of the building of this engine which gives an insight into some of the complexity involved and the building methods.

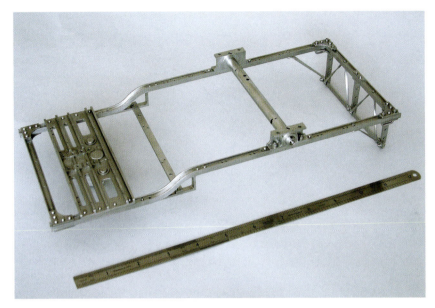
One of the engine frames. The curved frame sides are machined from solid metal.

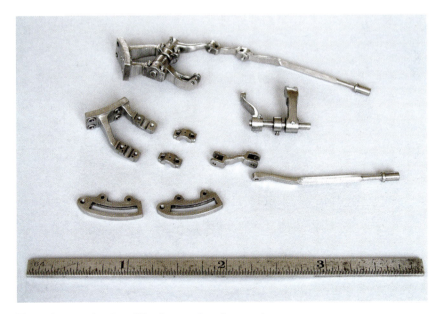
The valve mechanism (Stephenson's valve gear).

The drum comprises the internal ring gear, spoked rings, strengthening rings and the outer roller.

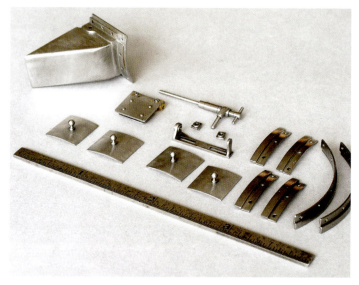

Various boiler attachments.

Weighshaft with lever attached and various brackets.

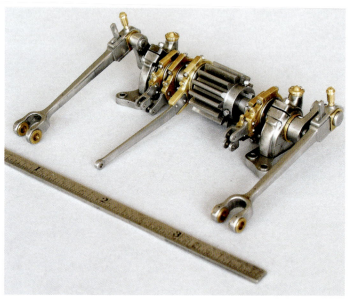

Assembled crankshaft.

THE ENGINE IN A WHEEL

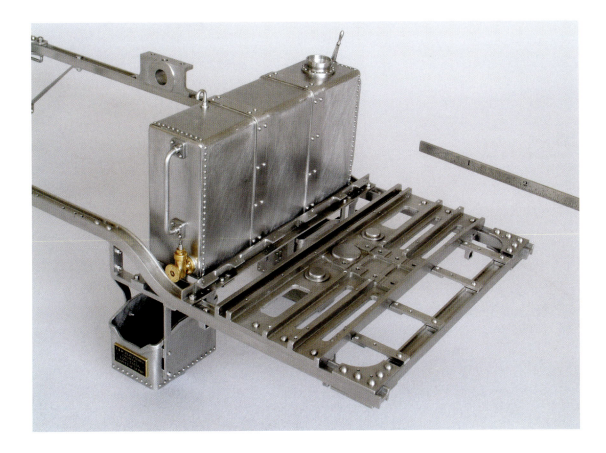

Tank and coal bunker fitted to frame.

Milling jig used to cut the curved frames.

Building models like the Blackburn also means making many special tools and jigs to produce various components. Pictured right and on the following pages are just some of them.

Jig for machining the large gear ring.

Some of the assorted tooling made to produce the Blackburn engine.

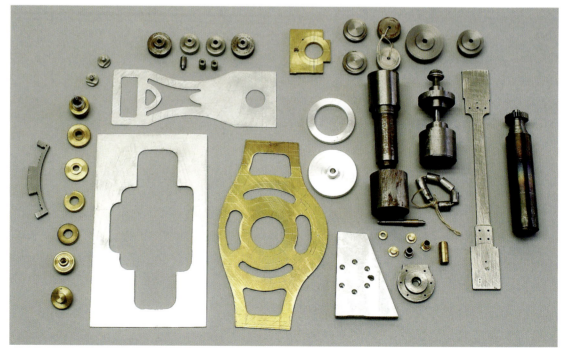

THE ENGINE IN A WHEEL

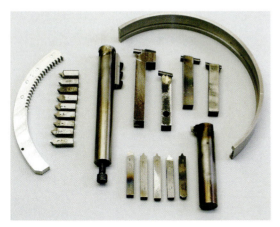

Tooling made to plane the internal ring gear teeth.

Once all the parts are made, the engine is fully built. Then it is broken down for painting, which takes many months. Photographs of the unpainted model show more clearly the details of construction and the quality of the work completed.

Some examples of unpainted work.

THE ENGINE IN A WHEEL

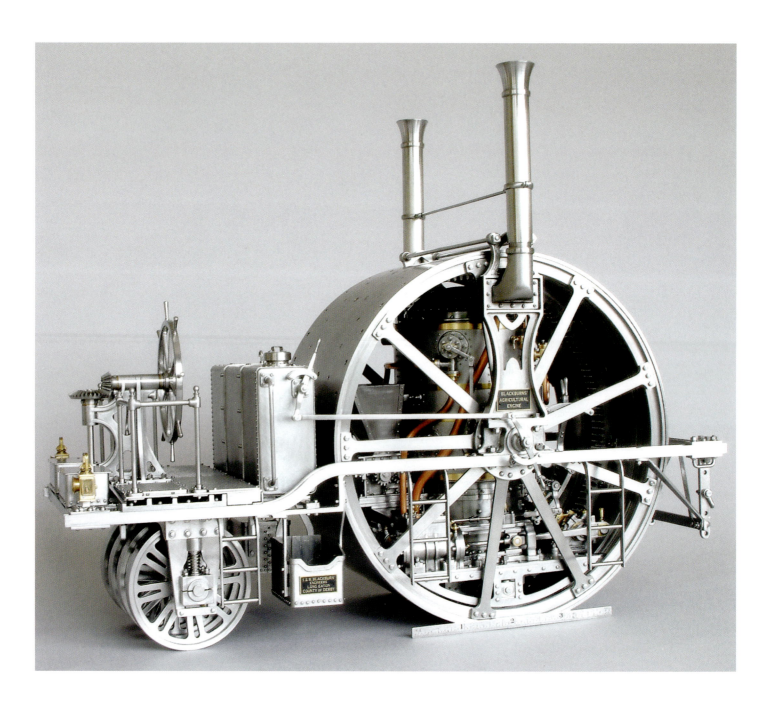

THE ENGINE IN A WHEEL

CHAPTER SIXTEEN

FILLING IN THE DETAIL

Blackburn agricultural engine 1863

Cherry Hill's second Blackburn agricultural engine, from 1863, won her a ninth Gold Medal, an eighth Bradbury Winter Award and a ninth Duke of Edinburgh Award. Both of the Blackburn designs were unusual in that the boiler was located inside a large drum which acted as the driving wheel. This arrangement presents a number of problems – getting coal and water into the boiler and the smoke out into the chimney are obvious ones. There are others.

This second engine, with a horizontal boiler, overcomes many of the problems of the first one, but still represents a real challenge to the modeller. The boiler and crankshaft are placed inside the drum, with the cylinders fixed to the front of the frame each side of the three guiding wheels. The boiler is connected to the frame. Flanged wheels are also fitted to the frame and these run on rails fixed to the inside of the open-sided drum. The flanged wheels are designed to keep a horizontal line, i.e. parallel to the ground on which it is travelling. The fire doors, ashpan doors, smoke box and chimney are at one end of the boiler; the combustion chamber at the other.

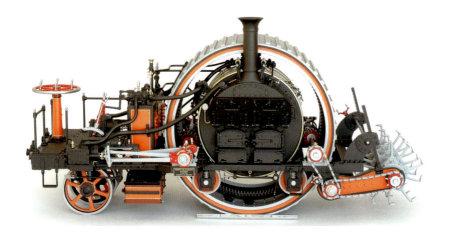

Blackburn agricultural engine 1863.

FILLING IN THE DETAIL

A challenge to the modeller.

The patent showed that a water tank was situated immediately in front of the drum, but no method of transferring water to the boiler was given. It also showed the position and design of the steering gear, the reversing lever and regulator lever, but no regulator unit. Front suspension was shown with differing positions of the brackets.

Otherwise, information on the patent drawings is incomplete or unclear in many respects. Cherry had to use her own ingenuity on some major components including the boiler, crankshaft, valve chest and eccentrics, front suspension and steering. Parts not shown at all include the regulator, pump, injector and water gauge. Choosing suitable designs and

Horizontal boiler is placed inside the drum.

locations for all these was part of the research and design processes.

The boiler was made from mild steel with copper firebox, flues and tubes. Fire doors, ashpan doors and the smoke box were located at the left-hand side of the boiler. The combustion chamber is on the right-hand side. A steam turret was fitted on top of the smoke box end, and the safety valves were fitted on top of the combustion chamber.

The drum was made from mild steel with bars riveted to the outside. The gear fitted inside the drum was designed in eight sections for the original; on the model it was made in one piece with pads spaced at each side of the line where the breaks would have been on the full-size engine.

The crankshaft design was modified as there was insufficient room between the frame and the drum for the position indicated on the patent. Valve chests were moved to the outside of the cylinders.

Other original work by Cherry included the cylinder exhaust, and fitting a blower. The front axle was supported at each end rather than between the wheels. The steering arrangement had a large-diameter steering wheel, with a pinion on the bottom of the shaft, and a spur gear.

The model engine is fitted with a digger/scarifier designed after researching other diggers of the time. The model digger has seven rings of tines fitted, each ring having

FILLING IN THE DETAIL

Smoke-tight boiler inspection doors.

Digger adjustment.

five equally spaced tines, alternatively advanced in successive rings. The height of the digger cylinder is adjustable by two racks and pinions operating on a shaft to a worm wheel. A shaft is fitted inside the drum to the rear of the boiler for the drive, and is driven by a dog-clutch operated pinion engaging with the large internal gear on the inside drum surface. A standard 'towing bracket' fitted for general agricultural work is removed in order to fit the digger cylinder.

The model is built to Cherry's usual scale of 1:16 or ¾in to 1ft. The drum is 6.86in over the bars, equivalent to 9ft 2in in full size.

CHERRY'S MODEL ENGINES

The boiler is made from steel with copper flues and tubes.

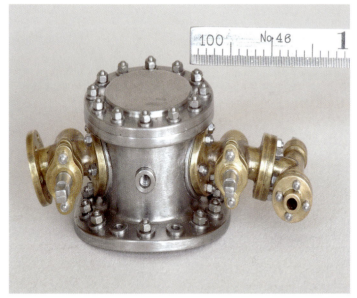

The boiler dome.

Ring gear teeth are planed.

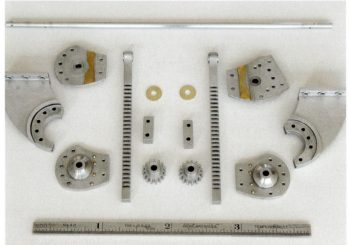

Digger drive and adjustment mechanism.

FILLING IN THE DETAIL

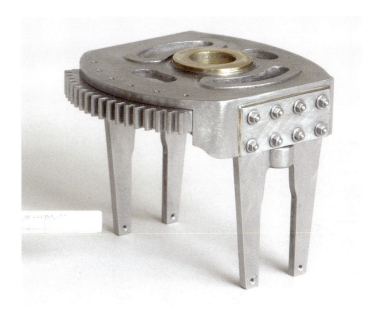

Steering frame.

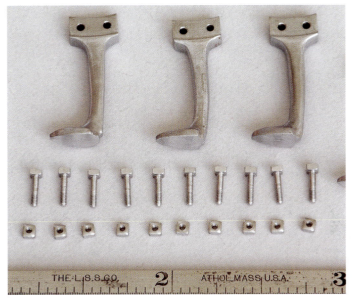

Digger tines.

Front suspension components.

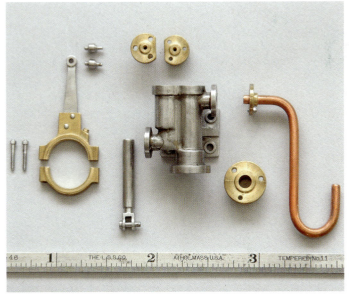

Fabricated pump body and drive parts.

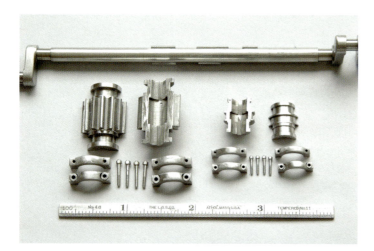

Crankshaft components.

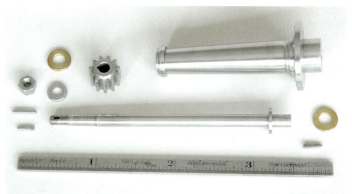

Steering column and shaft.

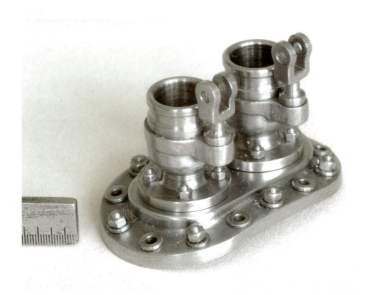

Safety valves under construction.

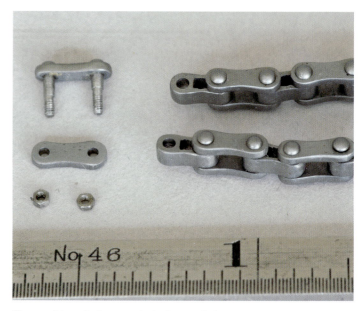

Digger drive chain – made in the workshop.

Drum width is 3¾in, equivalent to 5ft. Model operating speed is 60 rpm – a scale equivalent of 2 mph. The model's full speed of 180 rpm represents a full-size speed of 6.1 mph.

No castings are used in this model, or any other model Cherry has built in the previous forty years. Everything on the Blackburn is fabricated or machined from solid: nothing is bought in. Cherry made the spur gears, internal gears, worm, worm wheel and racks. Also the transmission chains and sprockets, small retaining chains, nameplates, chequer plates and all nuts, bolts and studs apart from a few fasteners used during fabrication, of which no trace remains to be seen. Every rivet is hand-made.

The paintwork is sublime and took five months – a lot for a model that is only about a foot long. Anything less than perfect is rejected and remade, however minor and inconsequential the fault may be. 'Bare metal' parts are given a finish to represent the type of iron that would have been used at the time of the original.

Even before the model proper was started, a mock-up was made to test the design. Many parts on the mock-up were completed to the standard of the finished model. However, no such parts are recycled on the final product; they are all made afresh.

In total the model took two years to research and make initial sketches and seven years to produce drawings and for the construction. The model contains around 7,400 parts and occupied some 7,500 hours of work over a nine-year period. On completion the model was given away, like all her other models; all the later ones being donated to the Institution of Mechanical Engineers of which Cherry was made a Companion.

So, does the model work? Yes. The 1863 Blackburn was tested successfully on compressed air around the workshop floor. Cherry tests all her models when completed. Some are working that would never have worked in the original Victorian design, including the 1863 Blackburn. Her designs are often sufficiently detailed for a full-sized version to be built, which would work better than the original.

CHAPTER SEVENTEEN

HOT COALS AND ICE

Nathaniel Grew ice locomotive

This model, is still at the early stages of construction at the time of writing. Once again it is unusual, to say the least. It will be something of a surprise when it is finished and goes on show, so this chapter is just to whet the appetite, with a few photographs.

Today the snowmobile is everyday travel for white winters. In the mid-nineteenth century, however, a Mr Nathaniel Grew came up with the idea of a steerable steam locomotive with sled runners at each end of the engine. Power was applied via a pair of driving wheels fitted with steel spikes. Unlike many of today's snowmobiles the ice locomotive was fitted with a cab when on the ice to keep the driver warm with help from the boiler.

Apart from cast-iron cylinders, wheel centres and eccentric sheaves, it was made from wrought iron and gunmetal.

The engine was exported to Moscow to transport goods across frozen rivers and lakes. Many a prayer must have been offered not to have to stay in one spot for very long.

It is interesting to look at some of the parts completed to see how Cherry works. As can be seen from the photographs, the engine frame is complete. One complete sled steering unit is complete; another is nearing the point of assembly. Those angled curved sled blade holders are not made from bent steel angle: to get them just so, they are milled from solid bar on a conventional milling machine (i.e. no CNC). Of course, at this point, Cherry still does not have a computer, although an iPad might be on the shopping list.

HOT COALS AND ICE

Consulting some design notes.

143

Underside of main frame.

Runners were cut from solid metal on a milling machine.

HOT COALS AND ICE

Boiler temporarily in place.

Front sledges.

Sledges fitted to steering unit.

MODELS OVERVIEW

Model no.	Based on	Description	Scale	Weight in case
1	Stuart No. 9 horizontal engine	From Stuart Turner castings	–	7.5 kg
2	Allchin traction engine	7 hp Royal Chester, 1925	1:16	10.5 kg
3	Vertical twin engine	Based on Stuart 10V design	–	2.8 kg
4	Merryweather fire engine	Twin cylinder engine, twin pumps, 1905	1:16	5.5 kg
5	Aveling & Porter 10-ton road roller	A.F. Compound, 1931	1:16	16 kg
6	Burrell scenic showman's engine	8 nhp compound Winston Churchill, 1922	1:12	30 kg
7	Wallis & Steevens road roller	Advance twin cylinder No. 8100, 1936	1:16	11 kg
8	Wallis & Steevens road roller	Simplicity single cylinder No. 8023, 1930	1:16	6.4 kg
9	James Taylor & Son traction engine	Taylor's Patent Steam Elephant, 1862	1:16	14.5 kg
10	Savage fairground engines	No. 6 centre engine 903, No. 4 organ engine, 1934	1:10	7.3 kg
11	W.F. Batho 25-ton road roller	Prototype, 1870	1:16	22.7 kg
12	Law & Downie traction engine	Walrus 1863 four-wheel drive, two axle steering	1:16	16 kg
13	E. Gellerat rouleau compresseur à vapeur	Twin engines Jacques IV, 1881	1:16	17.3 kg
14/15	Andrew Barclay, traction engine 1862–3	Tractor unit William and boring engine Alexander	1:16	20 kg
16	Gilletts & Allatt traction engine	Four-wheel drive and two axle steering	1:16	15.4 kg
17	Blackburn agricultural engine, 1857	Boiler and cylinders inside driving wheel	1:16	15.4 kg
18	Blackburn agricultural engine, 1863	Boiler and cylinders inside driving wheel	1:16	15.4 kg
19	Nathaniel Grew ice locomotive	Steerable locomotive on sled runners	1:16	–

Model no.	Building time	Awards
1	1956–7	Bronze Medal, 1964
2	1957–64	Silver Medal, 1964
3	Six months in 1996	Bronze Medal, 1968
4	1964–6	Silver Medal and Bradbury Winter Trophy, 1968
5	1966–9, approx. 2,500 hours	Championship Cup*, Crebbin Memorial Trophy, 1970, Duke of Edinburgh Award, 1971
6	1969–75, approx. 8,000 hours	Championship Cup*, Bradbury Winter Trophy, 1976
7	1975–9	Championship Cup* and Bradbury Winter Trophy, 1980
8	1975–9	Championship Cup* and Bradbury Winter Trophy, 1980, Duke of Edinburgh Award, 1981
9	1978–83, approx. 4,000 hours	Gold Medal, Aveling Barford Cup, 1984
10	1968 and 1982–4, approx. 2,500 hours	Gold Medal, 1985
11	1983–6, approx. 6,000 hours	Gold Medal, Aveling Barford Cup, 1987, Duke of Edinburgh Award, 1990
12	1986–90, approx. 6,500 hours	Gold Medal and Bradbury Winter Trophy, 1992, Duke of Edinburgh Award, 1993
13	1986–91, 1995, approx. 6,600 hours	Gold Medal and Bradbury Winter Trophy, 1996, Duke of Edinburgh Award, 1997
14/15	1988–94, approx. 9,000 hours	Gold Medal and Bradbury Winter Trophy, 1998, Duke of Edinburgh Award, 1999
16	1990–98 and 2001, approx. 6,500 hours	Gold Medal and Bradbury Winter Trophy, 2002, Duke of Edinburgh Award, 2003
17	1996–2003, 2005, approx. 6,500 hours	Gold Medal and Bradbury Winter Trophy, 2005, Duke of Edinburgh Award, 2007
18	Completed 2010, approx. 7,500 hours	Gold Medal and Bradbury Winter Trophy, 2010, Duke of Edinburgh Award, 2011
19	Still under construction, 2014	–

*The Championship Cup was the forerunner of the Gold Medal, awarded only once a year

INDEX

Allatt 11
Allchin Royal Chester 38–41
Andrew Barclay traction and boring engine 99–108
Aveling & Porter AF road roller 45–9
awards 10

Batho 25-ton road roller, 1870 78–86
Blackburn agricultural engine, 1857 20–23, 117–33
Blackburn agricultural engine, 1863 134–41
Bomford, Norris 15
Bonds O' Euston Road 18, 36
Burrell Showman's Engine 50–57

Centre engine No. 6 73–4
Crypton Synchrocheck 18

Duke of Edinburgh Award 8, 9, 10

Florida 15

Gellerat steam roller, 1881 94–8
Gilletts 11
Gilletts and Allatt traction engine, 1862 109–16
gold medals 9, 10

Hill, Ivor 14, 15
Hinds, George 11, 47
Hughes, Bill 40
Humber Special 15

Institution of Mechanical Engineers 9

Jones, Michael 9

Law & Downie road locomotive, 1863 87–93
Law, Ivan 8

Mays, Norman 9, 19
Merryweather fire engine 19, 41–4
Mock-ups 24
Model Engineer Exhibition 9, 41, 43
Model Engineer magazine 9, 14, 40, 43
motorcycles 14
Myford 15

Nathaniel Grew ice locomotive 142–6

Organ Engine No. 4 73, 75–7

painting the models 32
Pittler lathe 11, 13

Royce, Sir Henry 9

Savage fairground engines, 1934 73–7
Society of Model and Experimental Engineers 36, 37
sports cars 17
Stuart No. 9 36
Stuart 10V 41
Stuart Turner 36
symmetrical polyhedra 30

Taylor's Steam Elephant, 1862 66–72

Wallis & Steevens' Advance roller, 1936 61–5
Wallis & Steevens' Simplicity roller, 1930 58–60
workshop equipment, USA 28, 29
workshops 25–7